SOLOMON ACADEMY Distribution or replication of part of this page is prohibited

Solomon Academy

SHSAT & TJHSST
MATH WORKBOOK

by Yeon Rhee

13 Lessons with Key Summaries

Review of Essential Theorems and Properties for the Admissions Test

8 Full-Length Practice Tests

Detailed Solutions for All Questions

www.solomonacademy.net

SOLOMON ACADEMY Distribution or replication of any part of this page is prohibited. **LEGAL NOTICE**

Legal Notice

The New York City Specialized High Schools Admissions Test (SHSAT) and Thomas Jefferson High School for Science and Technology (TJHSST) Admissions Test was not involved in the production of this publication nor endorses this book.

Copyright © 2014 by Solomon Academy
Published by: Solomon Academy
First Edition
ISBN-13: 978-1497521773
ISBN-10: 1497521777

All rights reserved. This publication or any portion thereof may not be copied, replicated, distributed, or transmitted in any form or by any means whether electronically or mechanically whatsoever. It is illegal to produce derivative works from this publication, in whole or in part, without the prior written permission of the publisher and author.

www.solomonacademy.net

Acknowledgements

I wish to acknowledge my deepest appreciation to my wife, Sookyung, who has continuously given me wholehearted support, encouragement, and love. Without you, I could not have completed this book.

Thank you to my sons, Joshua and Jason, who have given me big smiles and inspiration. I love you all.

Thank you to my wonderful editor, Daniel Kwon, who has given me great advice and invaluable help.

About This Book

This book is designed to help you master the quantitative reasoning sections of the New York City Specialized High School Admissions Test (SHSAT) and Thomas Jefferson High School for Science and Technology (TJHSST) Admissions Test. The book contains 13 topic-specific summaries and 10 problems relevant to each section. Along with the topic-specific lessons, there are 8 full-length practice test with detailed solutions and explanations. It is recommended that you take Test 1 as a diagnostic test to understand your current level of expertise and in which area you need improvement. Afterwards, review the key lessons and essential theorems of the Thomas Jefferson Admissions Test and SHSAT. After completing the lessons, use the seven remaining practice tests to help improve your score and exhibit real test-taking conditions. There is no greater substitute than to study and practice.

Be sure to time yourself during the mathematic test with the appropriate time limit of 60 minutes. After completing any lessons or tests, immediately use the answer key and detailed solution to check your answers. Review all answers. Take the time to carefully read the explanations of problems you got incorrect. If you find yourself continually missing the same type of questions, look back at the topic summaries and review the theorems and examples in the lesson. Set a goal of improvement for each practice test.

About Author

Yeon Rhee obtained a Masters of Arts Degree in Statistics at Columbia University, NY. He served as the Mathematical Statistician at the Bureau of Labor Statistics, DC. He is the Head Academic Director at Solomon Academy due to his devotion to the community coupled with his passion for teaching. His mission is to help students of all confidence level excel in academia to build a strong foundation in character, knowledge, and wisdom.

About SHSAT and TJHSST Admissions Test

About SHSAT

The Specialized High Schools Admissions Test (SHSAT) are used by the Specialized High Schools in New York City: Bronx High School of Science, Brooklyn Latin School, Brooklyn Technical High School, High School of American Studies at Lehman College, High School for Math, Science and Engineering at City College, Queens High School for the Sciences at York College, Staten Island Technical High School, and Stuyvesant High School. Under contract to the New York City Department of Education, the SHSAT is developed and administered by American Guidance Service, a subsidiary of Pearson Education.

About Thomas Jefferson High School

Thomas Jefferson High School for Science and Technology (TJHSST) was established in 1985 with the collaboration of businesses and schools to improve education in science, mathematics, and technology. TJHSST is a Virginia state-chartered magnet school and one of 18 Virginia Governors Schools. It is ranked the best public high school in the United States of America. By providing students with a challenging learning environment, the rigorous coursework is focused on math, science, and technology in order to foster student success through innovative thinking based on ethical behavior.

About The Admissions Test

The New York City SHSAT and Thomas Jefferson High School for Science and Technology (TJHSST) Admissions Test scores are an opportunity for students to show their basic understand and ability to apply knowledge in math, reading, and writing obtained in academia. The test measures literacy and writing skills along with the quantitative section. Admissions boards are interested in critical thinking, logic-based reasoning, innovative problem-solving abilities, and communication skills. The SHSAT and TJHSST admissions Test score is one of many criteria that the admission boards use to consider an applicant; along with prior academic achievement, grade point average (GPA), intensity of rigorous course work, involvement in the community, extra-curricular activities, essays, and letters of recommendation.

This particular book is designed to help you master the quantitative reasoning section. The Mathematics section has 50 questions consisting of word problems and computation questions during an allotted 60 minutes. There is only one correct answer to each question and there is no penalty for guessing. Cell phones are prohibited at test sites and students are not allowed to bring any electronic devices including, but not limited to, calculators, computation aids, and dictionaries. Therefore, when taking practice tests and preparing, do not use a calculator. This book undergoes a summary of the key concepts and theorems needed to successfully face the mathematics section of the SHSAT and TJHSST admissions Test.

Contents

Lesson 1		Number Theory	9
Lesson 2		Simplifying Numerical Expressions	12
Lesson 3		Simplifying and Evaluating Algebraic Expressions	15
Lesson 4		Properties of Exponents	19
Lesson 5		Solving Equations and Word Problems	23
Lesson 6		Solving Inequalities and Compound Inequalities	27
Lesson 7		Fractions, Ratios, Rates, and Proportions	31
Lesson 8		Linear Equations	35
Lesson 9		Solving Systems of Linear Equations	39
Lesson 10		Classifying Angles	45
Lesson 11		Properties and Theorems of Triangles	50
Lesson 12		Patterns and Data Analysis	56
Lesson 13		Counting and Probability	61
Practice Test 1			68
Practice Test 1 Solutions			76
Practice Test 2			87
Practice Test 2 Solutions			96
Practice Test 3			107
Practice Test 3 Solutions			116
Practice Test 4			128
Practice Test 4 Solutions			136

SOLOMON ACADEMY

TABLE of CONTENTS

Practice Test 5	147
Practice Test 5 Solutions	155
Practice Test 6	167
Practice Test 6 Solutions	175
Practice Test 7	185
Practice Test 7 Solutions	193
Practice Test 8	206
Practice Test 8 Solutions	215

LESSON 1

Number Theory

Number theory is a branch of mathematics devoted to the study of integers. It involves analyzing relationships between integers such as factors, prime numbers, prime factorization, remainders, etc.

Factors

Factors are the numbers that you multiply to get another number. For instance, $10 = 1 \times 10$ and $10 = 2 \times 5$. Thus, the factors of 10 are 1, 2, 5, and 10.

Prime numbers

A prime number is a whole number that has only two factors: 1 and itself. The prime numbers less than 60 are 2, 3, 5, 7, 11, 13, 17, 19, 23, 29, 31, 37, 41, 43, 47, 53, and 59. It is worth noting that 2 is the first, smallest, and only even prime number among the prime numbers. A composite number is a number that has more than two factors. 0 and 1 are neither prime nor composite. For instance, 4 has three factors: 1, 2, and 4. Thus, 4 is a composite number.

Prime factorization

A prime factorization of a number is a process of writing the number as the product of all its prime factors. Since 12 can be written as $12 = 2 \times 2 \times 3$, the prime factorization of 12 is $2^2 \times 3$.

The prime factorization is useful when you count the total number of factors of a number. If a prime factorization of a number, n, is $n = 2^a \times 3^b \times 5^c$, the total number of factors of n is $(a+1) \times (b+1) \times (c+1)$. For instance, a prime factorization of 72 is $72 = 2^3 \times 3^2$. Thus, the total number of factors of 72 is $(3+1) \times (2+1) = 12$.

The total number of factors of a perfect square is always odd. For instance, the prime factorization of 9 is $9 = 3^2$. Thus, the total number of factors is $(2+1) = 3$.

Remainder

A remainder, r, is amount left over when a number, n, is divided by a divisor, p. When a number, n, is divided by p, the quotient is q and the remainder is r. Then, the number, n, can be expressed in terms of p, q, and r: $n = pq + r$. For instance, when 7 is divided by 2, the quotient is 3 and the remainder is 1. Thus, 7 can be expressed as $7 = 2 \times 3 + 1$.

When a number is divided by p, the possible values of the remainder are either $0, 1, 2, \cdots, p-2$, or $p-1$. For instance, if a number, n, is divided by 5, the possible values of the remainder are either 0, 1, 2, 3, or, 4.

SOLOMON ACADEMY — LESSON 1

Example: How many factors does 42 have?

The prime factorization of 42 is $42 = 2^1 \times 3^1 \times 7^1$. Thus, the total number of factors of 42 is $(1+1) \times (1+1) \times (1+1) = 8$. Or, you can find the total number factors of 42 by listing them out. Since $42 = 1 \times 42$, $42 = 2 \times 21$, $42 = 3 \times 14$, and $42 = 6 \times 7$, the total number of factors is 8: 1, 2, 3, 6, 7, 14, 21 and 42.

EXERCISES

1. Which of the following number has 6 factors?

   (A) 25   (B) 24   (C) 22
   (D) 16   (E) 12

2. Which of the following number is a prime number?

   (A) 53   (B) 51   (C) 39
   (D) 21   (E) 1

3. Which of the following number has an odd number of factors?

   (A) 11   (B) 28   (C) 36
   (D) 45   (E) 60

4. If the prime factorization of 200 is $2^a \times 5^b$, what is the value of $a + b$?

   (A) 4   (B) 5   (C) 6
   (D) 7   (E) 8

5. How many factors does 60 have?

   (A) 10   (B) 12   (C) 13
   (D) 16   (E) 18

6. Which of the following number has only two factors?

   (A) 15   (B) 36   (C) 43
   (D) 49   (E) 51

7. If a positive integer, n, is divided by k, the quotient is 7 and the remainder is 2. What is n in terms of k?

   (A) $2k + 7$ (B) $2k + 9$
   (C) $5k + 2$ (D) $5k + 7$
   (E) $7k + 2$

8. If the prime factorization of a number, k, is $2^3 \times 3^2$, what is the prime factorization of $3k$?

   (A) $2^4 \times 3^2$ (B) $2^3 \times 3^5$
   (C) $2^3 \times 3^4$ (D) $2^3 \times 3^3$
   (E) $2^2 \times 3^3$

9. Which of the following number is NOT a factor of 144?

   (A) 3   (B) 9   (C) 16
   (D) 24   (E) 27

10. If a positive integer, n, is divided by 4, the remainder is 3. What is the value of the remainder when $n + 3$ is divided by 4?

    (A) 2   (B) 3   (C) 4
    (D) 5   (E) 6

SOLOMON ACADEMY

Distribution or replication of any part of this page is prohibited.

LESSON 1

ANSWERS AND SOLUTIONS

1. (E)

   The prime factorization of 12 is $12 = 2^2 \times 3^1$. Thus, the total number of factors is $(2+1) \times (1+1) = 6$. Or, list the factors of 12: 1, 2, 3, 4, 6, and 12.

2. (A)

   Since 1 has only one factor, 1, it is not a prime number. 51, 39, and 21 are divisible by 3. Thus, they are not prime numbers. Only prime number in the answer choices is 53. Therefore, (A) is the correct answer.

3. (C)

   Any perfect squares have an odd number of factors. Since 36 is a perfect square, it has an odd number of factors. The prime factorization of 36 is $36 = 2^2 \times 3^2$. Thus, the total number of factors is $(2+1) \times (2+1) = 9$. Therefore, (C) is the correct answer.

4. (B)

   Since the prime factorization of 200 is $200 = 2^3 \times 5^2$, $a = 3$ and $b = 2$. Therefore, the value of $a + b = 3 + 2 = 5$.

5. (B)

   The prime factorization of 60 is $60 = 2^2 \times 3^1 \times 5^1$. Thus, the total number of factors is $(2+1) \times (1+1) \times (1+1) = 12$. Therefore, (B) is the correct answer.

6. (C)

   Any prime numbers have only two factors. Since 43 is a prime number, it has only two factors. Therefore, (C) is the correct answer.

7. (E)

   When a positive integer, n, is divided by k, the quotient is 7 and the remainder is 2. Thus, n can be expressed as $7k + 2$. Therefore, (E) is the correct answer.

8. (D)

   Since the prime factorization of k is $2^3 \times 3^2$, the prime factorization of $3k$ is $3k = 3 \times 2^3 \times 3^2 = 2^3 \times 3^3$. Therefore, (D) is the correct answer.

9. (E)

   144 is divisible by 3, 9, 16, and 24. However, it is not divisible by 27. Therefore, (E) is the correct answer.

10. (A)

    For simplicity, let n be 7 so that the remainder is 3 when 7 is divided by 4. Since $n + 3 = 10$, the remainder is 2 when 10 is divided by 4. Therefore, (A) is the correct answer.

www.solomonacademy.net

SOLOMON ACADEMY — LESSON 2

LESSON 2

Simplifying Numerical Expressions

To simplify numerical expressions, use the **order of operations** (PEMDAS)

- P: Parenthesis
- E: Exponent
- M: Multiplication
- D: Division
- A: Addition
- S: Subtraction

The order of operations suggests to first perform any calculations inside parentheses. Afterwards, evaluate any exponents. Next, perform all multiplications and divisions working from left to right. Finally, do additions and subtractions from left to right.

Example: Simplify $12(3-4)^2 \div 4 - 2$

$$\begin{aligned}
12(3-4)^2 \div 4 - 2 &= 12(-1)^2 \div 4 - 2 & \text{Simplify inside the parenthesis}\\
&= 12(-1)^2 \div 4 - 2 & \text{Evaluate the exponent}\\
&= 12 \div 4 - 2 & \text{Do multiplication and division from left to right}\\
&= 3 - 2 & \text{Do Subtraction}\\
&= 1
\end{aligned}$$

EXERCISES

1. $1 + 2 + 3 - 4 + 5 + 6 + 7 - 8 =$

   (A) 4 (B) 6 (C) 8 (D) 10 (E) 12

2. $6 \times 3 - 12 \div 2 =$

   (A) 12 (B) 9 (C) 6 (D) 3 (E) 1

3. Which of the following value is equal to $\frac{1}{2} \times 4 \times \frac{1}{3} \times 2 \times \frac{1}{4} \times 3$?

   (A) 1 (B) 2 (C) 3 (D) 4 (E) 5

4. Which of the following value is equal to the expression below?

   $$2(1+2)^2 + (2-6)$$

   (A) 5 (B) 10 (C) 14 (D) 24 (E) 32

5. Which of the following value is equal to the expression $(-1)^2 + (-1)^3 + (-1)^4$?

   (A) 1 (B) 2 (C) 3 (D) 4 (E) 5

www.solomonacademy.net

SOLOMON ACADEMY — Distribution or replication of any part of this page is prohibited. — **LESSON 2**

6. $\dfrac{10(3)^2 - 5 \times 4}{5(3+4)} =$

   (A) 1 (B) 2 (C) 3 (D) 4 (E) 5

7. Which of the following value is equal to the expression below?

   $$4\left(\sqrt{16} + \sqrt{25}\right) - 10^2 \div 4$$

   (A) 5 (B) 7 (C) 9
   (D) 11 (E) 13

8. What is the value of $\dfrac{3}{5}(3^2 + 4^2 + 5^2)$?

   (A) 30 (B) 40 (C) 50
   (D) 60 (E) 70

9. $10(2+3-4)^2 - 9(2+3-4)^2 =$

   (A) 0 (B) 1 (C) 2 (D) 3 (E) 4

10. $5|-2| - |4 - 2 \times 6| =$

    (A) 0 (B) 2 (C) 4 (D) 6 (E) 8

ANSWERS AND SOLUTIONS

1. (E)

   Since the expression has only additions and subtractions, simplify the expression from the left to right.

   $$1 + 2 + 3 - 4 + 5 + 6 + 7 - 8 = 12$$

2. (A)

   Do multiplication and division first. Then, add and subtract.

   $$6 \times 3 - 12 \div 2 = 18 - 6$$
   $$= 12$$

3. (A)

   Since the expression has only multiplications, rearrange the integers and fractions to simplify the expression easily.

   $$\frac{1}{2} \times 4 \times \frac{1}{3} \times 2 \times \frac{1}{4} \times 3 = \left(2 \times \frac{1}{2}\right) \times \left(3 \times \frac{1}{3}\right) \times \left(4 \times \frac{1}{4}\right)$$
   $$= 1 \times 1 \times 1$$
   $$= 1$$

4. (C)

   Use the order of operations (PEMDAS).

   $$2(1+2)^2 + (2-6) = 2(3)^2 - 4$$
   $$= 2 \times 9 - 4$$
   $$= 14$$

5. (A)

   $(-1)^2 = 1$, $(-1)^3 = -1$, and $(-1)^4 = 1$. Therefore, $(-1)^2 + (-1)^3 + (-1)^4 = 1$

SOLOMON ACADEMY

LESSON 2

Distribution or replication of any part of this page is prohibited.

6. **(B)**

    Use the order of operations (PEMDAS).

    $$\frac{10(3)^2 - 5 \times 4}{5(3+4)} = \frac{10(9) - (5 \times 4)}{5(7)}$$
    $$= \frac{90 - 20}{35}$$
    $$= 2$$

7. **(D)**

    Since $\sqrt{16} = 4$ and $\sqrt{25} = 5$,

    $$4(\sqrt{16} + \sqrt{25}) - 10^2 \div 4 = 4(4+5) - (10^2 \div 4)$$
    $$= 36 - 25$$
    $$= 11$$

8. **(A)**

    Simplify the expression inside the parenthesis: $(3^2 + 4^2 + 5^2) = 50$. Therefore,

    $$\frac{3}{5}(3^2 + 4^2 + 5^2) = \frac{3}{5} \times 50$$
    $$= 30$$

9. **(B)**

    Since the expression inside the parenthesis is $(2 + 3 - 4)^2 = 1$,

    $$10(2+3-4)^2 - 9(2+3-4)^2 = 10(1) - 9(1)$$
    $$= 1$$

10. **(B)**

    Since $|-2| = 2$ and $|4 - (2 \times 6)| = |-8| = 8$,

    $$5|-2| - |4 - 2 \times 6| = 5(2) - 8$$
    $$= 2$$

www.solomonacademy.net

LESSON 3

Simplifying and Evaluating Algebraic Expressions

Like terms are terms that have same variables and same exponents; only the coefficients may be different but can be the same. Knowing like terms is essential when you simplify algebraic expressions. For instance,

- $2x$ and $3x$: (Like terms)
- $2x$ and $3x^2$: (Not like terms since the two expressions have different exponents)
- 2 and 3: (Like terms)

Use the **distributive property** to expand an algebraic expression that has a parenthesis.

$$x(y+z) = x \times y + x \times z$$

To simplify an algebraic expression, expand the expression using the distributive property. Then group the like terms and simplify them. For instance,

$$\begin{aligned} 2(-x+2) + 3x + 5 &= -2x + 4 + 3x + 5 \quad &\text{Use distributive property to expand} \\ &= (-2x + 3x) + (4 + 5) \quad &\text{Group the like terms and simplify} \\ &= x + 9 \end{aligned}$$

To evaluate an algebraic expression, substitute the numerical value into the variable. When substituting a negative numerical value, make sure to use a **parenthesis** to avoid a mistake.

Example 1: Simplify the expression $\quad 3(x^2 - 2x + 3) - 2(x^2 - x + 2)$

$$\begin{aligned} 3(x^2 - 2x + 3) - 2(x^2 - x + 2) &= 3x^2 - 6x + 9 - 2x^2 + 2x - 4 \\ &= (3x^2 - 2x^2) + (-6x + 2x) + (9 - 4) \\ &= x^2 - 4x + 5 \end{aligned}$$

Example 2: Evaluate the expression $\quad -2x^2 + 3x \quad$ when $\quad x = -3$

$$\begin{aligned} -2x^2 + 3x &= -2(-3)^2 + 3(-3) \\ &= -2(9) - 9 \\ &= -27 \end{aligned}$$

SOLOMON ACADEMY — LESSON 3

EXERCISES

1. Which of the following expression equals $-(2-x)+2$?

   (A) x (B) $x-2$ (C) $x-4$
   (D) $2x$ (E) $3x$

2. Evaluate $(x-4)(x-1)$ when $x=3$

   (A) -2 (B) -1 (C) 0
   (D) 1 (E) 2

3. Simplify $3x - 4y + 5y - 2x$

   (A) $5x+9$ (B) $-x-y$ (C) $-x+y$
   (D) $x+y$ (E) $x-y$

4. If the length of a square is $2x-3$, what is the perimeter of the square?

   (A) $4x-6$ (B) $8x-12$ (C) x^2-2x
   (D) x^2-9 (E) $4x^2+9$

5. Evaluate $x^2 - yz + x + yz$ when $x=3$, $y=-2$, and $z=-3$.

   (A) 8 (B) 10 (C) 12
   (D) 14 (E) 16

6. Which of the following expression is equal to the sum of $4x+y$ and $2x+3y$ subtracted from $9x+5y$?

   (A) $x+3y$ (B) $3x+y$ (C) $3x-y$
   (D) $6x+4y$ (E) $5x+4y$

7. If $x=30$, evaluate the expression below.

   $$\frac{2x-10}{3} + \frac{3x-20}{3} + \frac{4x+30}{3}$$

   (A) 30 (B) 45 (C) 60
   (D) 75 (E) 90

8. What is the average of $9x+5$ and $5x+7$?

   (A) $6x+7$ (B) $7x+6$ (C) $7x+12$
   (D) $14x+6$ (E) $14x+12$

9. Evaluate $(\sqrt{x} - \sqrt{y})^2$ if $x=9$ and $y=25$

   (A) 256 (B) 64 (C) 16
   (D) 4 (E) 1

10. If Joshua is y years old now and his brother Jason is 7 years younger, what is Jason's age in 10 years from now?

    (A) $y-7$ (B) $y-3$ (C) $y+3$
    (D) $3y+3$ (E) $3y+7$

ANSWERS AND SOLUTIONS

1. (A)

   Use the distributive property to expand the expression inside the parenthesis.

   $$-(2-x)+2 = x-2+2$$
   $$= x$$

   Therefore, the expression $-(2-x)+2$ equals to x.

2. (A)

   In order to evaluate the expression, it is not necessary to expand the expression $(x-4)(x-1)$. Just substitute 3 for x in the expression.

   $$\begin{aligned}(x-4)(x-1) &= (3-4)(3-1) \qquad \text{Substitute 3 for } x \\ &= (-1)(2) \\ &= -2\end{aligned}$$

   Therefore, the value of $(x-4)(x-1)$ when $x=3$ is -2.

3. (D)

   Group the like terms and simplify them.

   $$\begin{aligned}3x - 4y + 5y - 2x &= (3x - 2x) + (-4y + 5y) \\ &= x + y\end{aligned}$$

4. (B)

   The length of the square is $2x - 3$. The perimeter of the square is four times the length of the square. Therefore,

   $$\text{Perimeter of square} = 4(2x - 3) = 8x - 12$$

5. (C)

   Before substituting 3 for x, -2 for y, and -3 for z in the expression, look at the expression carefully. You will notice that $-yz$ and yz are like terms that cancel each other out. Thus, the expression simplifies to $x^2 + x$. So, substitute 3 for x in $x^2 + x$.

   $$\begin{aligned}x^2 - yz + x + yz &= x^2 + x + (yz - yz) \\ &= x^2 + x \qquad \text{Substitute 3 for } x \\ &= (3)^2 + 3 \\ &= 12\end{aligned}$$

   Therefore, the value of $x^2 - yz + x + yz$ when $x = 3$, $y = -2$, and $z = -3$ is 12.

6. (B)

   The sum of $4x + y$ and $2x + 3y$ equals to $6x + 4y$, which is subtracted from $9x + 5y$.

   $$\begin{aligned}9x + 5y - (4x + y + 2x + 3y) &= 9x + 5y - (4x + 2x + y + 3y) \\ &= 9x + 5y - (6x + 4y) \\ &= 9x + 5y - 6x - 4y \\ &= (9x - 6x) + (5y - 4y) \\ &= 3x + y\end{aligned}$$

   Therefore, the expression equals to the sum of $4x + y$ and $2x + 3y$ which is subtracted from $9x + 5y$ is $3x + y$.

7. (E)

Before substituting 30 for x, simplify the expression by adding the three fractions with the same denominators.

$$\frac{2x-10}{3} + \frac{3x-20}{3} + \frac{4x+30}{3} = \frac{2x-10+3x-20+4x+30}{3}$$
$$= \frac{2x+3x+4x-10-20+30}{3}$$
$$= \frac{9x}{3}$$
$$= 3x$$

Thus, the expression simplifies to $3x$. Therefore, the value of the expression is $3x = 3(30) = 90$.

8. (B)

To evaluate an average of two expressions, divide the sum of the two expressions by two.

$$\text{Average} = \frac{\text{Sum of two expressions}}{2}$$
$$= \frac{9x+5+5x+7}{2}$$
$$= \frac{14x+12}{2}$$
$$= 7x+6$$

Therefore, the average of $9x+5$ and $5x+7$ is $7x+6$.

9. (D)

Since $\sqrt{9} = 3$ and $\sqrt{25} = 5$, $\sqrt{9} - \sqrt{25} = -2$. Therefore,

$$(\sqrt{x} - \sqrt{y})^2 = (\sqrt{9} - \sqrt{25})^2 \qquad \text{Substitute 9 for } x \text{ and 25 for } y$$
$$= (3-5)^2$$
$$= (-2)^2$$
$$= 4$$

10. (C)

Jason is 7 years younger than his brother, Joshua, who is y years old now. So, Jason's current age is $y - 7$ years old. Therefore, 10 years from now, Jason's age will be $y - 7 + 10 = y + 3$ years old.

SOLOMON ACADEMY — LESSON 4

LESSON 4
Properties of Exponents

In the expression 2^4, 2 is the base, 4 is the exponent, and 2^4 is the power. Exponents represent how many times the base is multiplied by. $2^4 = 2 \times 2 \times 2 \times 2$. The table below shows a summary of the properties of exponents.

Properties of Exponents	Example
1. $a^m \cdot a^n = a^{m+n}$	1. $2^4 \cdot 2^6 = 2^{10}$
2. $\frac{a^m}{a^n} = a^{m-n}$	2. $\frac{2^{10}}{2^3} = 2^{10-3} = 2^7$
3. $(a^m)^n = a^{mn} = (a^n)^m$	3. $(2^3)^4 = 2^{12} = (2^4)^3$
4. $a^0 = 1$	4. $(-2)^0 = 1,\ (3)^0 = 1,\ (100)^0 = 1$
5. $a^{-1} = \frac{1}{a}$	5. $2^{-1} = \frac{1}{2}$
6. $a^{\frac{1}{n}} = \sqrt[n]{a}$	6. $2^{\frac{1}{2}} = \sqrt{2},\quad x^{\frac{1}{3}} = \sqrt[3]{x}$
7. $(ab)^n = a^n b^n$	7. $(2 \cdot 3)^6 = 2^6 \cdot 3^6,\quad (2x)^2 = 2^2 x^2$
8. $\left(\frac{a}{b}\right)^n = \frac{a^n}{b^n}$	8. $\left(\frac{2}{x}\right)^3 = \frac{2^3}{x^3}$

To solve an exponential equation, make sure that expressions on both sides have the same base. If the expressions have the same base, then exponents on both sides are the same.

$$a^x = a^y \implies x = y \qquad \text{Example: } 2^x = 2^3 \implies x = 3$$

Example 1: Simplify $\left(\frac{1}{8}\right)^{\frac{2}{3}} \left(\frac{1}{8}\right)^{-\frac{2}{3}}$

Since the two expressions have the same base, $\frac{1}{8}$, you can combine both expressions using the first exponent property shown on the table above.

$$\left(\frac{1}{8}\right)^{\frac{2}{3}} \left(\frac{1}{8}\right)^{-\frac{2}{3}} = \left(\frac{1}{8}\right)^{\frac{2}{3}-\frac{2}{3}} = \left(\frac{1}{8}\right)^0 = 1$$

Example 2: Solve $3^{x+1} = 27$

$$3^{x+1} = 3^3 \qquad \text{Since both expressions have the same base}$$
$$x + 1 = 3$$
$$x = 2$$

SOLOMON ACADEMY — LESSON 4

EXERCISES

1. Simplify $\dfrac{x^3 \cdot x^4}{x^2}$

    (A) x^6 (B) x^5 (C) x^4
    (D) x^3 (E) x^2

2. What is the value of $2 \cdot 2^0 \cdot 2^{-1}$?

    (A) 4   (B) 3   (C) 2   (D) 1   (E) 0

3. If $x = \tfrac{1}{2}$, what is the value of x^{-2} ?

    (A) 1   (B) 2   (C) 4   (D) 6   (E) 8

4. If $x^2 = 3$, what is the value of x^6 ?

    (A) 6   (B) 8   (C) 9
    (D) 15   (E) 27

5. If $2^x = 64$, what is the value of x^2 ?

    (A) 64   (B) 36   (C) 25
    (D) 12   (E) 8

6. If $10^x = a$, what is 100^x in terms of a ?

    (A) $2a$ (B) $5a$ (C) $10a$
    (D) a^2 (E) 2^a

7. If $\dfrac{x}{y} = 3$, what is the value of $\dfrac{y^3}{x^3}$?

    (A) $\dfrac{1}{27}$ (B) $\dfrac{1}{18}$ (C) $\dfrac{1}{9}$
    (D) 9   (E) 27

8. If $x^2 = 2$ and $y^3 = 3$, what is the value of $(xy)^6$?

    (A) 6   (B) 8   (C) 48
    (D) 54   (E) 72

9. If $4^{3x+1} = 256$, what is the value of x ?

    (A) 0   (B) 1   (C) 2   (D) 3   (E) 4

10. If two positive integers, x and y, satisfy the following equation $24 \times 18 = 2^x 3^y$, what is the sum of the values of x and y ?

    (A) 3   (B) 4   (C) 7   (D) 12   (E) 81

ANSWERS AND SOLUTIONS

1. (B)
    Since the expressions have the same base, x, use the properties of exponents. Therefore,
    $$\dfrac{x^3 \cdot x^4}{x^2} = x^{3+4-2} = x^5$$

2. (D)
    Since $2^0 = 1$ and $2^{-1} = \tfrac{1}{2}$,
    $$2 \cdot 2^0 \cdot 2^{-1} = 2 \cdot 1 \cdot \dfrac{1}{2} = 1$$

SOLOMON ACADEMY — LESSON 4

Distribution or replication of any part of this page is prohibited.

3. (C)

If $x = \frac{1}{2}$, $x^2 = \frac{1}{4}$. Since $x^{-2} = (x^2)^{-1} = \frac{1}{x^2}$,

$$x^{-2} = \frac{1}{x^2} = \frac{1}{\frac{1}{4}} = 4$$

Therefore, the value of x^{-2} is 4.

4. (E)

Since $x^2 = 3$, $x^6 = (x^2)^3 = 3^3 = 27$.

5. (B)

$2^6 = 64$. Thus, $x = 6$. Therefore, the value of x^2 is $6^2 = 36$.

6. (D)

Since $10^x = a$,

$$100^x = (10^2)^x = 10^{2x} = (10^x)^2 = a^2$$

Therefore, 100^x in terms of a is a^2.

7. (A)

Since $\frac{x}{y} = 3$, $\frac{y}{x} = \frac{1}{3}$. Thus,

$$\frac{y^3}{x^3} = \left(\frac{y}{x}\right)^3 = \left(\frac{1}{3}\right)^3 = \frac{1}{27}$$

Therefore, the value of $\frac{y^3}{x^3}$ is $\frac{1}{27}$.

8. (E)

First, expand $(xy)^6$ by using the properties of exponents. Afterwards, substitute 2 for x^2 and 3 for y^3 in the expression.

$$\begin{aligned}(xy)^6 &= x^6 y^6 \\ &= (x^2)^3 (y^3)^2 \qquad \text{Substitute 2 for } x^2 \text{ and 3 for } y^3 \\ &= 8(9) \\ &= 72\end{aligned}$$

Therefore, the value of $(xy)^6$ is 72.

SOLOMON ACADEMY   Distribution or replication of any part of this page is prohibited.   **LESSON 4**

9. (B)

   To solve an exponential equation, both expressions must have the same base. Change 256 to 4^4 and then solve the equation.

   $$4^{3x+1} = 256$$
   $$4^{3x+1} = 4^4 \quad \text{Since both sides have the same base}$$
   $$3x + 1 = 4$$
   $$x = 1$$

   Therefore, the value of x is 1.

10. (C)

    Find the prime factorization of 24 and 18 separately: $24 = 2^3 \times 3$, and $18 = 2 \times 3^2$.

    $$24 \times 18 = 2^x 3^y$$
    $$2^3 \times 3 \times 2 \times 3^2 = 2^x 3^y$$
    $$2^4 \times 3^3 = 2^x 3^y$$

    Thus, $x = 4$ and $y = 3$. Therefore, the sum of the values of $x + y$ is $4 + 3 = 7$.

www.solomonacademy.net

LESSON 5

Solving Equations and Word Problems

Solving an equation is finding the value of the variable that makes the equation true. In order to solve an equation, use the rule called SADMEP with inverse operations (SADMEP is the reverse order of the order of operations, PEMDAS). Inverse operations are the operations that cancel each other. Addition and subtraction, and multiplication and division are good examples.

SADMEP suggests to first cancel subtraction or addition. Then, cancel division or multiplication next by applying corresponding inverse operation. Below is an example that shows you how to solve $2x - 1 = 5$, which involves subtraction and multiplication.

$$2x - 1 = 5$$
$$+1 = +1$$
$$2x = 6$$
$$x = 3$$

$\checkmark \quad \checkmark$
$S\ A\ D\ M\ E\ P$
Addition to cancel substraction
Division to cancel multiplication

Solving word problems involve translating verbal phrases into mathematical equations. The table below summarizes the guidelines.

Verbal Phrase	Expression
A number	x
Is	$=$
Of	$\times$
Percent	0.01 or $\frac{1}{100}$
The sum of x and y	$x + y$
Three more than twice a number	$2x + 3$
The difference of x and y	$x - y$
3 is subtracted from a number	$x - 3$
4 less than a number	$x - 4$
A number decreased by 5	$x - 5$
6 less a number	$6 - x$
The product of x and y	xy
6 times a number	$6x$
The quotient of x and y	$\frac{x}{y}$
A number divided by 9	$\frac{x}{9}$

SOLOMON ACADEMY — LESSON 5

Example: 5 more than the quotient of x and 3 is 14. What is the number?

$$\frac{x}{3} + 5 = 14$$
$$-5 = -5$$
$$\frac{x}{3} = 9$$
$$x = 27$$

$S\ A\ D\ M\ E\ P$

Subtraction to cancel addition

Multiplication to cancel division

EXERCISES

1. If $5x - 3 = 7$, what is the value of $2x$?

   (A) 0 (B) 1 (C) 2 (D) 3 (E) 4

2. If $3(x - 2) = 2(x - 2)$, then $x =$

   (A) 2 (B) 3 (C) 4 (D) 6 (E) 8

3. If $5x - (x + 6) = 18$, what is the value of $x + 6$?

   (A) 4 (B) 8 (C) 12 (D) 16 (E) 20

4. If $x^2 - x - 2 = x^2 - 2x - 3$, then $x =$

   (A) -3 (B) -2 (C) -1 (D) 1 (E) 3

5. If $2x$ subtracted from 6 equals 9 less than x, what is the value of x?

   (A) 6 (B) 5 (C) 4 (D) 3 (E) 2

6. If $2(x+y) = 6$, what is the value of $3x + 3y$?

   (A) 5 (B) 7 (C) 8 (D) 9 (E) 12

7. If $4xy - 6 = 2x + 3y$ and $y = 2$, then what is the value of x?

   (A) 2 (B) 3 (C) 4 (D) 5 (E) 6

8. If two thirds of the sum of $6x$ and 9 equals $2x$ less 8, what is the value of x?

   (A) -8 (B) -7 (C) -5 (D) 3 (E) 7

9. If $x = \frac{2y}{a}$, which of the following expression equals $6a$?

   (A) $\frac{3x}{y}$ (B) $\frac{6x}{y}$ (C) $\frac{6y}{x}$ (D) $\frac{12x}{y}$ (E) $\frac{12y}{x}$

10. Joshua has $300 in his savings account. If he saves $200 per week, in how many weeks will he have saved a total amount of $2100 in his savings account?

    (A) 5 (B) 6 (C) 7 (D) 8 (E) 9

SOLOMON ACADEMY
Distribution or replication of any part of this page is prohibited.

LESSON 5

ANSWERS AND SOLUTIONS

1. (E)

$$
\begin{aligned}
5x - 3 &= 7 \\
+3 &= +3 \\
5x &= 10 \\
x &= 2
\end{aligned}
$$

$\checkmark\checkmark$
$S\ A\ D\ M\ E\ P$
Addition to cancel substraction
Division to cancel multiplication

Therefore, the value of $2x$ is $2(2) = 4$.

2. (A)

Expand the expressions inside the parenthesis on the left and right side by using the distributive property.

$$
\begin{aligned}
3(x-2) &= 2(x-2) \qquad \text{Use the distributive property}\\
3x - 6 &= 2x - 4 \\
3x - 2x &= -4 + 6 \\
x &= 2
\end{aligned}
$$

3. (C)

Simplify the expression on the left side and solve the equation.

$$
\begin{aligned}
5x - (x+6) &= 18 \\
5x - x - 6 &= 18 \\
4x &= 24 \\
x &= 6
\end{aligned}
$$

Therefore, $x + 6 = 12$.

4. (C)

Cancel out x^2 on each side and solve for x.

$$
\begin{aligned}
x^2 - x - 2 &= x^2 - 2x - 3 \qquad \text{Subtract } x^2 \text{ from each side}\\
-x - 2 &= -2x - 3 \\
2x - x &= -3 + 2 \\
x &= -1
\end{aligned}
$$

5. (B)

$2x$ subtracted from 6 can be expressed as $6 - 2x$. 9 less than x can be expressed as $x - 9$.

$$
\begin{aligned}
6 - 2x &= x - 9 \\
-3x &= -15 \\
x &= 5
\end{aligned}
$$

SOLOMON ACADEMY — LESSON 5

6. (D)

You can not solve for the value of x and y separately since you only have one equation with two variables. Instead, multiply the equation by $\frac{3}{2}$ and find the value of $3x + 3y$.

$$\frac{3}{2} \times 2(x+y) = \frac{3}{2} \times 6 \qquad \text{Multiply both sides by } \frac{3}{2}$$
$$3(x+y) = 9$$
$$3x + 3y = 9$$

7. (A)

Substituting 2 for y simplifies the equation to $8x - 6 = 2x + 6$.

$$4xy - 6 = 2x + 3y \qquad \text{Substitute 2 for } y$$
$$8x - 6 = 2x + 6$$
$$6x = 12$$
$$x = 2$$

8. (B)

Two thirds of the sum of $6x$ and 9 can be expressed as $\frac{2}{3}(6x + 9)$ and $2x$ less 8 can be expressed as $2x - 8$.

$$\frac{2}{3}(6x + 9) = 2x - 8 \qquad \text{Use the distributive property}$$
$$4x + 6 = 2x - 8$$
$$2x = -14$$
$$x = -7$$

9. (E)

From the equation $x = \frac{2y}{a}$, write a in terms of y and x. Afterwards, evaluate $6a$.

$$x = \frac{2y}{a} \qquad \text{Multiply both sides by } a$$
$$ax = 2y \qquad \text{Divide both sides by } x$$
$$a = \frac{2y}{x} \qquad \text{Multiply both sides by 6}$$
$$6a = \frac{12y}{x}$$

10. (E)

Define x as the number of weeks Joshua needs to save the total amount of $2100. Since Joshua will save $200 per week, the amounts that Joshua will save in x weeks will be $200x$. Thus,

$$200x + 300 = 2100$$
$$200x = 1800$$
$$x = 9$$

Therefore, Joshua needs 9 weeks to save the total amount of $2100.

SOLOMON ACADEMY — LESSON 6

LESSON 6

Solving Inequalities and Compound Inequalities

Solving Inequalities
Solving an inequality is exactly the same as solving an equation. To solve an inequality, use SADMEP (Reverse order of the PEMDAS). In most cases, the inequality symbol remains unchanged. However, there are only two cases in which the inequality symbol must be reversed. The first case is when you multiply or divide each side by a negative number. The second case is when you take a reciprocal of each side. For instance,

Case 1
$2 < 3$
$-2 > -3$

Case 2
$2 < 3$
$\dfrac{1}{2} > \dfrac{1}{3}$

Solving Compound Inequalities
There are two types of compound inequalities: **And** compound inequality and **Or** compound inequality.

And Compound Inequality: $-5 \leq 2x - 1 \leq 7$

Or Compound Inequality: $x - 4 < -3$ or $2x + 1 > 7$

Below shows how to solve each type of compound inequality.

$-5 \leq 2x - 1 \leq 7$ And compound inequality
$+1 \leq \quad +1 \leq +1$ Add 1 to each side
$-4 \leq 2x \leq 8$ Divide each side by 2
$-2 \leq x \leq 4$

Thus, x is greater than or equal to -2 <u>and</u> less than or equal to 4

$x - 4 < -3$ or $2x + 1 > 7$ Or compound inequality
$x < 1$ or $x > 3$

Thus, x is less than 1 <u>or</u> greater than 3

Example: Solve $-3x + 2 > x + 10$

$-3x + 2 > x + 10$ Subtract x from each side
$-4x + 2 > 10$ Subtract 2 from each side
$-4x > 8$ Divide each side by -4
$x < -2$ Reverse the inequality symbol

SOLOMON ACADEMY
LESSON 6

EXERCISES

1. Solve the inequality $2x + 1 < 5$

    (A) $x < 2$ (B) $x < -2$ (C) $x > 2$
    (D) $x > -2$ (E) $x < -3$

2. Solve the inequality $-2 \leq 3x + 1 \leq 10$

    (A) $-3 \leq x \leq 3$ (B) $-1 \leq x \leq 1$
    (C) $0 \leq x \leq 2$ (D) $-3 \leq x \leq 1$
    (E) $-1 \leq x \leq 3$

3. Solve the inequality $-4x - 8 > 12$

    (A) $x > -5$ (B) $x < -5$ (C) $x > 5$
    (D) $x < 5$ (E) $x < 1$

4. What is the solution to the following inequality $3x + 1 < -5$ or $2x - 1 > 7$?

    (A) $x < -2$ or $x > 4$
    (B) $x < -2$ or $x > -4$
    (C) $x < 2$ or $x > 4$
    (D) $-2 < x < 4$
    (E) $2 < x < 4$

5. Solve $3(2x - 4) < 2(x + 4)$

    (A) $x < 3$ (B) $x > 4$ (C) $x < 5$
    (D) $x > 6$ (E) $x < 7$

6. How many positive integer values of x satisfy $-2(x - 8) > x - 2$?

    (A) 1   (B) 2   (C) 3
    (D) 4   (E) 5

7. If the solution to $4 - 2x > x + 16$ is $x < a$, what is the value of a ?

    (A) 6   (B) 4   (C) 2
    (D) -4 (E) -6

8. Solve $-10 \leq -3x - 4 < 5$

    (A) $-3 \leq x < 2$ (B) $-3 < x \leq 2$
    (C) $-2 \leq x < 3$ (D) $-2 < x \leq 3$
    (E) $2 \leq x < -3$

9. If 4 less than a number is less than 4 and greater than -3, find the number.

    (A) $0 < x < 7$ (B) $1 < x < 7$
    (C) $1 < x < 8$ (D) $0 < x < 8$
    (E) $2 < x < 9$

10. If 3 more than twice a number is at most 11 and at least 5, find the number.

    (A) $2 < x < 8$ (B) $2 \leq x \leq 8$
    (C) $1 \leq x < 4$ (D) $1 < x \leq 4$
    (E) $1 \leq x \leq 4$

ANSWERS AND SOLUTIONS

1. (A)

    $2x + 1 < 5$ Subtract 1 from each side
    $2x < 4$ Divide each side by 2
    $x < 2$

SOLOMON ACADEMY — LESSON 6

2. (E)

$$-2 \leq 3x+1 \leq 10 \quad \text{Subtract 1 from each side}$$
$$-3 \leq 3x \leq 9 \quad \text{Divide each side by 3}$$
$$-1 \leq x \leq 3$$

3. (B)

$$-4x - 8 > 12 \quad \text{Add 8 to each side}$$
$$-4x > 20 \quad \text{Divide each side by } -4$$
$$x < -5 \quad \text{Reverse the inequality symbol}$$

4. (A)

$$3x + 1 < -5 \quad \text{or} \quad 2x - 1 > 7$$
$$3x < -6 \quad \text{or} \quad 2x > 8$$
$$x < -2 \quad \text{or} \quad x > 4$$

5. (C)

$$3(2x - 4) < 2(x + 4) \quad \text{Expand each side using the distributive property}$$
$$6x - 12 < 2x + 8 \quad \text{Subtract } 2x \text{ from each side}$$
$$4x - 12 < 8 \quad \text{Add 12 to each side}$$
$$4x < 20 \quad \text{Divide each side by 4}$$
$$x < 5$$

6. (E)

$$-2(x - 8) > x - 2 \quad \text{Expand left side using the distributive property}$$
$$-2x + 16 > x - 2 \quad \text{Subtract } x \text{ from each side}$$
$$-3x + 16 > -2 \quad \text{Subtract 16 from each side}$$
$$-3x > -18 \quad \text{Divide each side by } -3 \text{ and reverse the inequality symbol}$$
$$x < 6$$

The positive integer values of x for which $x < 6$ are 1, 2, 3, 4, and 5. Therefore, there are 5 positive integer values of x that satisfy the original inequality.

7. (D)

$$4 - 2x > x + 16 \quad \text{Subtract } x \text{ from each side}$$
$$4 - 3x > 16 \quad \text{Subtract 4 from each side}$$
$$-3x > 12 \quad \text{Divide each side by } -3 \text{ and reverse the inequality symbol}$$
$$x < -4$$

Therefore, the value of a is -4.

SOLOMON ACADEMY   Distribution or replication of any part of this page is prohibited.   **LESSON 6**

8. (B)

$$-10 \le -3x - 4 < 5 \quad \text{Add 4 to each side}$$
$$-6 \le -3x < 9 \quad \text{Divide each side by } -3 \text{ and reverse the inequality symbols}$$
$$2 \ge x > -3 \quad \text{Rearrange the inequality}$$
$$-3 < x \le 2$$

9. (C)

Translate the verbal phrases into an **And compound** inequality. Let x be the number. Then, 4 less than a number can be expressed as $x - 4$.

$$-3 < x - 4 < 4 \quad \text{Add 4 to each side}$$
$$1 < x < 8$$

10. (E)

At most means $\le$ and **at least** means $\ge$. Let x be the number. Then, 3 more than a twice a number can be expressed as $2x + 3$.

$$5 \le 2x + 3 \le 11 \quad \text{Subtract 3 from each side}$$
$$2 \le 2x \le 8 \quad \text{Divide each side by 2}$$
$$1 \le x \le 4$$

LESSON 7

Fractions, Ratios, Rates, and Proportions

A **fraction** represents a part of a whole. In the fraction $\frac{4}{5}$, the numerator 4 means that the fraction represents 4 equal parts, and the denominator 5 means that 5 parts make up a whole.

A **ratio** is a fraction that compares two quantities measured in the same units. The ratio of a to b can be written as $a : b$ or $\frac{a}{b}$. If the ratio of a number of apples to that of oranges in a store is $3 : 4$ or $\frac{3}{4}$, it means that there are 3 apples to every 4 oranges in the store.

A **rate** is a ratio that compares two quantities measured in different units. A rate is usually expressed as a unit rate. A unit rate is a rate per one unit of a given quantity. The rate of a per b can be written as $\frac{a}{b}$. If a car travels 100 miles in 2 hours, the car travels at a rate of 50 miles per hour.

A **proportion** is an equation that states that two ratios are equal. A proportion can be written as

$$a : b = c : d \quad \text{or} \quad \frac{a}{b} = \frac{c}{d}$$

The proportion above reads a is to b as c is to d. To solve the value of a variable in a proportion, use the cross product property and then solve for the variable. For instance,

$$\frac{x}{2} = \frac{6}{3} \qquad \text{Cross product property}$$
$$3x = 2 \times 6$$
$$x = 4$$

Example: Simplify $\frac{x}{2} + \frac{x}{3}$

$$\frac{x}{2} + \frac{x}{3} = \frac{3x}{6} + \frac{2x}{6} \qquad \text{Common denominator is 6}$$
$$= \frac{5x}{6}$$

EXERCISES

1. Simplify $\frac{2x-3y}{y}$

   (A) $\frac{2x}{y} - 3$ (B) $\frac{2x}{y} + 2$ (C) $\frac{2}{xy} - 3$
   (D) $\frac{2}{xy} + 2$ (E) $2x - 2y$

2. Two-thirds of students in a class are girls. If one-half of the girls wear glasses, what fractional part of students are girls who wear glasses?

   (A) $\frac{1}{8}$ (B) $\frac{1}{6}$ (C) $\frac{1}{5}$
   (D) $\frac{1}{4}$ (E) $\frac{1}{3}$

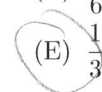

SOLOMON ACADEMY — LESSON 7

3. If the ratio of a to b is $2 : 3$ and the ratio of c to b is $3 : 4$, which of the following is equal to the ratio of a to c?

   (A) $\dfrac{1}{2}$ (B) $\dfrac{2}{3}$ (C) $\dfrac{3}{4}$ (D) $\dfrac{8}{9}$ (E) $\dfrac{9}{10}$

4. If a student can type 120 words in 3 minutes, at this rate, how many words can she type in 5 minutes?

   (A) 240 (B) 200 (C) 180 (D) 160 (E) 120

5. A 45-inch string is cut into two pieces. If the ratio of the longer piece to the shorter piece is $3 : 2$, what is the length of the shorter piece?

   (A) 9 (B) 12 (C) 18 (D) 27 (E) 36

6. If $2x - 3y = 0$, what is $\dfrac{x}{y}$?

   (A) $\dfrac{3}{2}$ (B) $\dfrac{2}{3}$ (C) $\dfrac{5}{3}$ (D) $\dfrac{3}{5}$ (E) $\dfrac{5}{2}$

7. $\dfrac{x}{2} + \dfrac{x}{3} + \dfrac{x}{4} = 26$, $x =$

   (A) 12 (B) 16 (C) 18 (D) 20 (E) 24

8. If a car travels at a rate of 48 miles per hour, how many miles does it travel in 40 minutes?

   (A) 28 (B) 32 (C) 36 (D) 40 (E) 42

9. 10 people who build at the same rate can frame a house in 6 days. What fractional part of a house can 4 people frame a house in 3 days?

   (A) $\dfrac{1}{2}$ (B) $\dfrac{1}{3}$ (C) $\dfrac{1}{4}$ (D) $\dfrac{1}{5}$ (E) $\dfrac{1}{6}$

10. If Joshua can wash x cars in y hours, how many cars does he wash in z hours?

    (A) $\dfrac{xy}{z}$ (B) $\dfrac{yz}{x}$ (C) $\dfrac{xz}{y}$ (D) $\dfrac{x}{yz}$ (E) $\dfrac{z}{xy}$

ANSWERS AND SOLUTIONS

1. (A)

$$\dfrac{2x - 3y}{y} = \dfrac{2x}{y} - \dfrac{3y}{y} = \dfrac{2x}{y} - 3$$

2. (E)

Let x be the total number of students in the class. Two-thirds of the students are girls. Thus, the number of girls can be expressed as $\dfrac{2}{3}x$. Since one-half of the girls wear glasses, the number of girls who wear glasses can be expressed as $\dfrac{1}{2}(\dfrac{2}{3}x)$ or $\dfrac{1}{3}x$. Therefore, one-third of the students are girls who wear glasses.

SOLOMON ACADEMY — LESSON 7

3. (D)

Since $\frac{a}{b} = \frac{2}{3}$ and $\frac{c}{b} = \frac{3}{4}$, the ratio $\frac{a}{c}$ can be calculated using the two given ratios.

$$\frac{a}{c} = \frac{a}{b} \times \frac{b}{c} \qquad \left(\text{Since } \frac{c}{b} = \frac{3}{4} \Longrightarrow \frac{b}{c} = \frac{4}{3}\right)$$
$$= \frac{2}{3} \times \frac{4}{3}$$
$$= \frac{8}{9}$$

4. (B)

If a student can type 120 words in 3 minutes, she can type $120 \div 3 = 40$ words in 1 minute. Therefore, she can type $40 \times 5 = 200$ words in 5 minutes.

5. (C)

The ratio of the longer piece to the shorter piece is $3 : 2$. Instead of using 3 for the longer piece and 2 for the shorter piece directly, multiply the ratio $3 : 2$ by x so that a new ratio is $3x : 2x$. Now, $3x$ represents the length of the longer piece and $2x$ represents the length of the shorter piece. If you add the longer and shorter pieces together, the sum of these lengths is equal to the length of the original string, 45 inches. Thus,

$$3x + 2x = 45$$
$$5x = 45$$
$$x = 9$$

Therefore, the length of the shorter piece is $2x = 2(9) = 18$ inches.

6. (A)

$$2x - 3y = 0 \qquad \text{Add } 3y \text{ to each side}$$
$$2x = 3y \qquad \text{Divide each side by 2}$$
$$x = \frac{3}{2}y \qquad \text{Divide each side by } y$$
$$\frac{x}{y} = \frac{3}{2}$$

7. (E)

The least common multiple of 2, 3, and 4 is 12. Multiply each side by 12 to eliminate fractions.

$$12 \times \left(\frac{x}{2} + \frac{x}{3} + \frac{x}{4}\right) = 26 \times 12 \qquad \text{Use the distributive property}$$
$$6x + 4x + 3x = 26(12)$$
$$13x = 26(12)$$
$$x = \frac{26(12)}{13}$$
$$x = 24$$

Therefore, the value of x is 24.

SOLOMON ACADEMY — LESSON 7

Distribution or replication of any part of this page is prohibited.

8. (B)

48 miles per hour means that the car travels 48 miles in one hour. There are 60 minutes in one hour. Set up a proportion in terms of miles and minutes.

$$48_{\text{miles}} : 60_{\text{minutes}} = x_{\text{miles}} : 40_{\text{minutes}}$$
$$\frac{48}{60} = \frac{x}{40} \quad \text{Use cross product property}$$
$$60x = 48 \times 40$$
$$x = 32$$

Therefore, the car travels 32 miles in 40 minutes.

9. (D)

Let's define work as the number of people times the number of days. The work required to frame a house is $10_{\text{people}} \times 6_{\text{days}} = 60_{\text{people} \times \text{days}}$. If 4 people frame in 3 days, the work they finish is $4_{\text{people}} \times 3_{\text{days}} = 12_{\text{people} \times \text{days}}$. Thus,

$$\text{A fractional part of a house} = \frac{12_{\text{people} \times \text{days}}}{60_{\text{people} \times \text{days}}}$$
$$= \frac{1}{5}$$

Therefore, 4 people can frame $\frac{1}{5}$ of the house in 3 days.

10. (C)

Let's set up a proportion and solve for w.

$$x_{\text{cars}} : y_{\text{hours}} = w_{\text{cars}} : z_{\text{hours}}$$
$$\frac{x}{y} = \frac{w}{z} \quad \text{Use cross product property}$$
$$yw = xz$$
$$w = \frac{xz}{y}$$

Therefore, Joshua can wash $\frac{xz}{y}$ cars in z hours.

LESSON 8
Linear Equations

The **slope**, m, of a line is a number that describes the steepness of the line. The larger the absolute value of the slope, $|m|$, the steeper the line is (closer to y-axis). If a line passes through the points (x_1, y_1) and (x_2, y_2), the slope m is defined as

$$m = \frac{\text{Rise}}{\text{Run}} = \frac{y_2 - y_1}{x_2 - x_1}$$

If the points (x_1, y_1) and (x_2, y_2) are given, the following formulas are useful in solving TJHSST and SHSAT math problems.

Midpoint Formula: $\left(\dfrac{x_1 + x_2}{2}, \dfrac{y_1 + y_2}{2}\right)$

Distance Formula: $D = \sqrt{(x_2 - x_1)^2 + (y_2 - y_1)^2}$

An equation of a line can be written in three different forms.

1. **Slope-intercept form:** $y = mx + b$, where m is slope and b is y-intercept.

2. **Point-slope form:** If the slope of a line is m and the line passes through the point (x_1, y_1),

$$y - y_1 = m(x - x_1)$$

3. **Standard form:** $Ax + By = C$, where A, B, and C are integers.

Below classifies the lines by slope.

- Lines that rise from left to right have positive slope.
- Lines that fall from left to right have negative slope.
- Horizontal lines have zero slope (example: $y = 2$).
- Vertical lines have undefined slope (example: $x = 2$).
- Parallel lines have the same slope.
- Perpendicular lines have negative reciprocal slopes (product of the slopes equals -1).

The x**-intercept** of a line is a point where the line crosses x-axis. The y**-intercept** of a line is a point where the line crosses y-axis.

To find the x-intercept of a line	$\implies$	Substitute 0 for y and solve for x
To find the y-intercept of a line	$\implies$	Substitute 0 for x and solve for y

SOLOMON ACADEMY — Distribution or replication of any part of this page is prohibited. **LESSON 8**

Example: If the slope of a line is 3 and the y-intercept is -4, write an equation of the line.

In slope-intercept form, $y = mx + b$, m represents the slope and b represents the y-intercept. Thus, $m = 2$ and $b = -4$. Therefore, the equation of the line is $y = 2x - 4$.

EXERCISES

1. If a line passes through the points $(-2, 3)$ and $(1, 9)$, what is the slope of the line?

    (A) -3 (B) -2 (C) -1
    (D) $\frac{1}{3}$ (E) 2

2. What are the x and y coordinates of the midpoint between $(6, 5)$ and $(-2, 1)$?

    (A) $(4, 3)$ (B) $(2, 4)$ (C) $(4, 2)$
    (D) $(3, 2)$ (E) $(2, 3)$

3. What is the distance between $(-1, -3)$ and $(4, 9)$?

    (A) 13 (B) 12 (C) 10
    (D) 8 (E) 5

4. Which of the following line has a slope of zero?

    (A) $y = 2x - 3$ (B) $y = -2x + 3$
    (C) $y = -2$ (D) $x = -2$
    (E) $x + y = 0$

5. If a point $(1, a)$ lies on the line $y = -3x + 4$, what is the value of a?

    (A) -1 (B) 0 (C) 1
    (D) 2 (E) 3

6. If the equation of the line is $2x + 3y = 6$, what is the x-intercept of the line?

    (A) -6 (B) -3 (C) 2
    (D) 3 (E) 6

7. Which of the following line is parallel to $y = 4x + 1$?

    (A) $y = 4x + 2$ (B) $y = -4x + 2$
    (C) $y = \frac{1}{4}x + 2$ (D) $y = -\frac{1}{4}x + 2$
    (E) $y = -\frac{1}{4}x - 2$

8. What is the equation of the line that is parallel to $y = \frac{1}{2}x + 3$ and passes $(4, 7)$?

    (A) $y = \frac{1}{2}x - 5$ (B) $y = \frac{1}{2}x + 5$
    (C) $y = -\frac{1}{2}x + 5$ (D) $y = -\frac{1}{2}x - 3$
    (E) $y = -2x + 5$

9. Which of the following line is perpendicular to $y = \frac{1}{3}x - 2$?

    (A) $-x + 3y = 6$ (B) $-x - 3y = 6$
    (C) $-2x + y = 4$ (D) $3x - y = 4$
    (E) $3x + y = 4$

10. What is the equation of the perpendicular bisector of a line segment connected by $(1, 5)$ and $(5, 3)$?

    (A) $y = 2x + 2$ (B) $y = 2x - 2$
    (C) $y = \frac{1}{2}x + 2$ (D) $y = \frac{1}{2}x - 2$
    (E) $y = -\frac{1}{2}x + 2$

SOLOMON ACADEMY — LESSON 8

ANSWERS AND SOLUTIONS

1. (E)
$$\text{Slope} = \frac{y_2 - y_1}{x_2 - x_1} = \frac{9 - 3}{1 - (-2)} = \frac{6}{3} = 2$$

2. (E)
$$\text{Midpoint} = \left(\frac{x_1 + x_2}{2}, \frac{y_1 + y_2}{2}\right) = \left(\frac{6 + (-2)}{2}, \frac{5 + 1}{2}\right) = (2, 3)$$

3. (A)
$$\begin{aligned}\text{Distance} &= \sqrt{(x_2 - x_1)^2 + (y_2 - y_1)^2} \\ &= \sqrt{(4 - (-1))^2 + (9 - (-3))^2} \\ &= \sqrt{5^2 + 12^2} \\ &= 13\end{aligned}$$

   Therefore, the distance between $(-1, -3)$ and $(4, 9)$ is 13.

4. (C)

   Horizontal lines have a slope of zero. Any horizontal lines can be written as $y = k$, where k=constant. Therefore, $y = -2$ is the answer.

5. (C)

   Since point, $(1, a)$, is on the line, $(1, a)$ is the solution to the equation $y = -3x + 4$. Substitute 1 for x and a for y in the equation and solve for a.

$$\begin{aligned} y &= -3x + 4 &&\text{Substitute 1 for } x \text{ and } a \text{ for } y \\ a &= -3(1) + 4 \\ a &= 1 \end{aligned}$$

   Therefore, the value of a is 1.

6. (D)

   To find the x-intercept of the line $2x + 3y = 6$, substitute 0 for y in the equation and solve for x.

$$\begin{aligned} 2x + 3y &= 6 &&\text{Substitute 0 for } y \\ 2x + 3(0) &= 6 \\ x &= 3 \end{aligned}$$

   Therefore, the x-intercept of the line is 3.

7. (A)

   Two lines are parallel if they have the same slope. Thus, the slope of the parallel line must be 4. Since the only equation of the line that has the slope of 4 is $y = 4x + 2$ in the answer choices, (A) is the correct answer.

SOLOMON ACADEMY
Distribution or replication of any part of this page is prohibited.
LESSON 8

8. (B)

   Two parallel lines have the same slope. Thus, the slope of the parallel line is $\frac{1}{2}$. Start with the slope-intercept form, $y = mx + b = \frac{1}{2}x + b$. Since the point $(4, 7)$ is on the line, $(4, 7)$ is a solution to the equation $y = \frac{1}{2}x + b$. Substitute 4 for x and 7 for y in the equation and then solve for b.

   $$y = \frac{1}{2}x + b \qquad \text{Substitute 4 for } x \text{ and 7 for } y$$
   $$7 = \frac{1}{2}(4) + b \qquad \text{Solve for } b$$
   $$b = 5$$

   Therefore, the equation of the parallel line that passes through $(4, 7)$ is $y = \frac{1}{2}x + 5$.

9. (E)

   The slope of the perpendicular line to $y = \frac{1}{3}x - 2$ is -3. Each equation in the answer choices is written in standard form. Rewrite each equation of the line in slope-intercept form and choose the equation of the line that has the slope of -3.

   $$\begin{aligned}
   \text{(A)} \quad &-x + 3y = 6 &\implies& \quad y = \frac{1}{3}x + 2 \\
   \text{(B)} \quad &-x - 3y = 6 &\implies& \quad y = -\frac{1}{3}x - 2 \\
   \text{(C)} \quad &-2x + y = 4 &\implies& \quad y = 2x + 4 \\
   \text{(D)} \quad &3x - y = 4 &\implies& \quad y = 3x - 4 \\
   \text{(E)} \quad &3x + y = 4 &\implies& \quad y = -3x + 4
   \end{aligned}$$

   Therefore, (E) is the correct answer.

10. (B)

    The slope of the line segment connected by $(1, 5)$ and $(5, 3)$ is

    $$\text{Slope} = \frac{3 - 5}{5 - 1} = -\frac{1}{2}$$

    The midpoint between $(1, 5)$ and $(5, 3)$ is

    $$\text{Midpoint} = \left(\frac{1 + 5}{2}, \frac{5 + 3}{2}\right) = (3, 4)$$

    The slope of the perpendicular bisector is the negative reciprocal of $-\frac{1}{2}$, or 2. The equation of the perpendicular bisector in slope-intercept form is $y = 2x + b$. Since the perpendicular bisector passes through the midpoint of the line segment, $(3, 4)$ is the solution to the equation $y = 2x + b$.

    $$y = 2x + b \qquad \text{Substitute 3 for } x \text{ and 4 for } y$$
    $$4 = 2(3) + b$$
    $$b = -2$$

    Therefore, the equation of the perpendicular bisector is $y = 2x - 2$.

LESSON 9

Solving Systems of Linear Equations

A system means more than one. A linear equation represents a line. Thus, a **system of linear equations** represent more than one line. Below is an example of a system of linear equations.

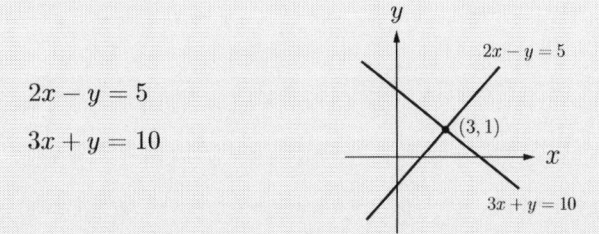

$$2x - y = 5$$
$$3x + y = 10$$

A solution to a system of linear equations is an ordered pair (x, y) that satisfies each equation in the system. In other words, a solution to a system of linear equation is an intersection point that lies on both lines. In the figure above, $(3, 1)$ is an ordered pair that satisfies each equation,

$$2x - y = 5 \implies 2(3) - 1 = 5$$
$$3x + y = 10 \implies 3(3) + 1 = 10$$

and is the intersection point of both lines.

Solving a system of linear equations means finding the x and y coordinates of the intersection point of the lines. There are two methods to solve a system of linear equations: **substitution** and **linear combinations**.

1. Substitution method
In the example above, write y in terms of x in the first equation. $2x - y = 5 \implies y = 2x - 5$. Substitute $2x - 5$ for y in the second equation.

$$3x + y = 10 \implies 3x + (2x - 5) = 10$$
$$5x - 5 = 10$$
$$x = 3 \implies y = 2x - 5 = 2(3) - 5 = 1$$

The solution to the system using the substitute method is $(3, 1)$.

2. Linear combinations method
In the example above, the coefficient of the y variable in each equation is opposite. Thus, adding the two equations eliminates the y variables. Then, solve for x.

$$2x - y = 15$$
$$\underline{3x + y = 10} \quad \text{Add two equations}$$
$$5x = 15$$
$$x = 3$$

SOLOMON ACADEMY — LESSON 9

Distribution or replication of any part of this page is prohibited.

Substitute 3 for x in the first equation and solve for y.

$$2x - y = 5 \implies 2(3) - y = 5 \implies y = 1$$

The solution to the system using the linear combinations method is $(3, 1)$.

Example: Solve the system of linear equations below.

$$y = x - 4$$
$$2x + y = 2$$

Solve the system of equations using the substitution method.

$$2x + y = 2 \quad \text{Substitute } y = x - 4 \text{ for } y$$
$$2x + (x - 4) = 2$$
$$3x - 4 = 2$$
$$x = 2 \implies y = x - 4 = (2) - 4 = -2$$

Therefore, the solution to the system is $(2, -2)$.

EXERCISES

1. If $3x + 2y = 5$ and $5x - 2y = 3$, what is the value of $x + 3$?

   (A) 1 (B) 2 (C) 3
   (D) 4 (E) 5

2. If $-x + 3y = 16$ and $y = 2x - 3$, what is the value of y?

   (A) 3 (B) 4 (C) 5
   (D) 6 (E) 7

3. If $x - y = 6$ and $2x + 4y = 9$, what is the value of $2x + 2y$?

   (A) 8   (B) 10 (C) 12
   (D) 15 (E) 18

4. If $-2x + 2y = 12$ and $\frac{x}{2} = \frac{y}{6}$, what is the value of x?

   (A) 2 (B) 3 (C) 4
   (D) 5 (E) 6

$$3x - y = m$$
$$-2x + 5y = n$$

5. If $x = 3$ and $y = 2$ are solutions to the system of equations above, what is the value of $m + n$?

   (A) 11 (B) 9 (C) 7
   (D) 5   (E) 3

6. If the two lines, $y = 2x$ and $y = 6 - x$ intersect, what are the x and y coordinates of the intersection point?

   (A) $(4, 8)$ (B) $(4, 2)$ (C) $(3, 6)$
   (D) $(3, 2)$ (E) $(2, 4)$

7. If $x - y = 1$ and $x + y = 7$, what is the value of $3x - 2y$?

   (A) 6 (B) 5 (C) 4
   (D) 3 (E) 2

www.solomonacademy.net

8. If $3x - 4y = -1$ and $-4x + 3y = 6$, what is the solution to the system of equations?

   (A) $(-3, -2)$ (B) $(-2, -3)$
   (C) $(3, -2)$ (D) $(3, 2)$
   (E) $(2, 3)$

9. Joshua and Jason saved $1000 together. Joshua saved $100 more than twice the amount that Jason saved. How much did Jason save?

   (A) $400   (B) $350   (C) $300
   (D) $250   (E) $200

10. A store sells desks and chairs. A store makes a profit of $15 per desk and $8 per chair. If the store sold a total of 23 desks and chairs and made the total profit of $240, how many chairs did the store sell?

    (A) 5   (B) 8   (C) 10
    (D) 12   (E) 15

ANSWERS AND SOLUTIONS

1. (D)

   Use the linear combinations method.

   $$\begin{aligned} 3x + 2y &= 5 \\ 5x - 2y &= 3 \\ \hline 8x &= 8 \\ x &= 1 \end{aligned}$$ Add two equations

   Therefore, $x + 3 = 4$.

2. (E)

   Use the substitution method.

   $$\begin{aligned} -x + 3y &= 16 \\ -x + 3(2x - 3) &= 16 \\ 5x - 9 &= 16 \quad \Longrightarrow \quad x = 5 \end{aligned}$$ Substitute $2x - 3$ for y

   Therefore, $y = 2x - 3 = 2(5) - 3 = 7$.

3. (B)

   Add the two equations and divide the result by $\frac{2}{3}$ to find the value of $2x + 2y$.

   $$\begin{aligned} x - y &= 6 \\ 2x + 4y &= 9 \\ \hline 3x + 3y &= 15 \\ 2x + 2y &= 10 \end{aligned}$$ Add two equations \
   Multiply each side by $\frac{2}{3}$

   Therefore, the value of $2x + 2y = 10$.

SOLOMON ACADEMY — LESSON 9

4. (B)

Multiply each side of the equation $\frac{x}{2} = \frac{y}{6}$ by 6 to obtain $y = 3x$. Then, use the substitution method.

$$-2x + 2y = 12 \quad \text{Substitute } 3x \text{ for } y$$
$$-2x + 2(3x) = 12$$
$$4x = 12$$
$$x = 3$$

Therefore, the value of x is 3.

5. (A)

Substitute 3 for x and 2 for y in the first equation to solve for m.

$$3x - y = m \quad \text{Substitute 3 for } x \text{ and 2 for } y$$
$$3(3) - 2 = m \quad \text{Solve for } m$$
$$m = 7$$

Then, substitute 3 for x and 2 for y in the second equation to solve for n.

$$-2x + 5y = n \quad \text{Substitute 3 for } x \text{ and 2 for } y$$
$$-2(3) + 5(2) = n \quad \text{Solve for } n$$
$$n = 4$$

Thus, $n = 4$. Therefore, the value of $m + n = 11$.

6. (E)

In order to find the intersection point that lies on both lines $y = 2x$ and $y = 6 - x$, use the substitution method.

$$y = 6 - x \quad \text{Substitute } 2x \text{ for } y$$
$$2x = 6 - x$$
$$3x = 6$$
$$x = 2$$

Thus, $x = 2$ and $y = 2x = 2(2) = 4$. Therefore, the x and y coordinates of the intersection point is $(2, 4)$.

7. (A)

Use the linear combinations method.

$$x - y = 1$$
$$\underline{x + y = 7} \quad \text{Add two equations}$$
$$2x = 8$$
$$x = 4$$

Since $x = 4$, substitute 4 for x in the first equation to solve for y.

$$x - y = 1 \qquad \text{Substitute 4 for } x$$
$$4 - y = 1 \qquad \text{Solve for } y$$
$$y = 3$$

Thus, $x = 4$ and $y = 3$. Therefore, the value of $3x - 2y = 3(4) - 2(3) = 6$.

8. (A)

Use the linear combinations method. Since the coefficients of x in the first and second equation are 3 and -4, find the least common multiple (LCM) of 3 and 4, which is 12. Thus, multiply the first equation by 4, and multiply the second equation by 3 to obtain the same coefficient of x, 12.

$$3x - 4y = -1 \quad \xrightarrow{\text{Multiply by 4}} \quad 12x - 16y = -4$$
$$-4x + 3y = 6 \quad \xrightarrow{\text{Multiply by 3}} \quad -12x + 9y = 18$$

Add two equations to eliminate x variables.

$$12x - 16y = -4$$
$$\underline{-12x + 9y = 18} \qquad \text{Add two equations}$$
$$-7y = 14$$
$$y = -2$$

Substitute -2 for y in the first equation and solve for x.

$$3x - 4y = -1 \qquad \text{Substitute } -2 \text{ for } y$$
$$3x - 4(-2) = -1$$
$$3x = -9$$
$$x = -3$$

Thus, $x = -3$ and $y = -2$. Therefore, the solution to the system of equations is $(-3, -2)$.

9. (C)

Let x be the amount that Jason saved. Since Joshua saved \$100 more than twice the amount that Jason saved, $2x + 100$ represents the amount that Joshua saved. Joshua and Jason saved \$1000 together. Thus, the sum of x and $2x + 100$ equals 1000.

$$x + 2x + 100 = 1000$$
$$3x + 100 = 1000$$
$$3x = 900$$
$$x = 300$$

Therefore, the amount that Jason saved, x, is \$300.

SOLOMON ACADEMY — Distribution or replication of any part of this page is prohibited. **LESSON 9**

10. (E)

Let's define x as the number of desks that the store sold and y as the number of chairs that the store sold. Set a system of equations using the x and y variables. First, set up the first equation in terms of a total number of desks and chairs that the store sold. The store sold a total of 23 desks and chairs: $x + y = 23$. Then, set up the second equation in terms of $240 profit that the store made after selling x numbers of desks and y numbers of chairs: $15x + 8y = 240$. Next, multiply each side of the first equation by -15.

$$x + y = 23 \xrightarrow{\text{Multiply by } -15} -15x - 15y = -345$$
$$15x + 8y = 240$$

Use the linear combinations method.

$$\begin{aligned} -15x - 15y &= -345 \\ \underline{15x + 8y} &= \underline{240} \qquad \text{Add two equations} \\ -7y &= -105 \\ y &= 15 \end{aligned}$$

Therefore, the number of chairs that the store sold, y, is 15.

LESSON 10

Classifying Angles

An angle is formed by two rays and is measured in degrees (°). The angle A is expressed as $\angle A$ and the measure of the angle A is expressed as $m\angle A$.

Two angles, A and B, that have the same measure are called **congruent angles**. They are expressed as $\angle A \cong \angle B$.

Angles are classified by their measures.

- Acute angle is less than 90°.

- Right angle is 90°.

- Obtuse angle is greater than 90°.

- Straight angle is 180°.

- Vertical angles are formed by intersecting two lines. Vertical angles are congruent. In the figure below, $\angle 1$ and $\angle 3$, and $\angle 2$ and $\angle 4$ are vertical angles.

- Complementary angles are two angles whose sum of their measures is 90°. In the figure below, $\angle 5$ and $\angle 6$ are complementary angles.

- Supplementary angles are two angles whose sum of their measures is 180°. In the figure below, $\angle 7$ and $\angle 8$ are supplementary angles.

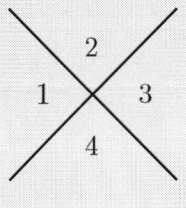

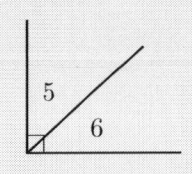

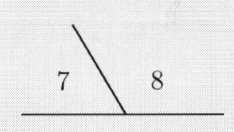

Vertical angles   Complementary angles   Supplementary angles

Example: Two angles are complementary. If one angle measures $x + 10$ and the other angle measures $2x + 5$, what is the value of x?

Two angles are complementary angles if the sum of their measures is 90°.

$$x + 10 + 2x + 5 = 90$$
$$3x + 15 = 90$$
$$3x = 75$$
$$x = 25$$

SOLOMON ACADEMY — LESSON 10

When two parallel lines are cut by a third line called the transversal, the following angles are formed.

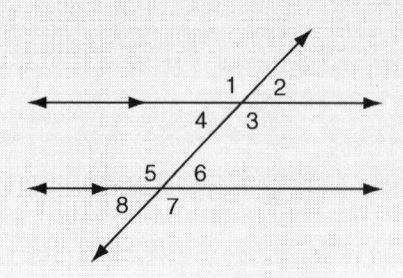

- Corresponding angles are congruent: $\angle 1 \cong \angle 5$, $\angle 4 \cong \angle 8$, $\angle 2 \cong \angle 6$, and $\angle 3 \cong \angle 7$.
- Alternate interior angles are congruent: $\angle 4 \cong \angle 6$, and $\angle 3 \cong \angle 5$.
- Alternate exterior angles are congruent: $\angle 1 \cong \angle 7$, and $\angle 2 \cong \angle 8$.
- Consecutive angles are supplementary: $\angle 4$ and $\angle 5$, and $\angle 3$ and $\angle 6$ are supplementary. In other words, $m\angle 4 + m\angle 5 = 180°$, and $m\angle 3 + m\angle 6 = 180°$.

EXERCISES

1. In the figure below, two lines intersect to form two pairs of vertical angles. What is the value of x?

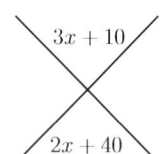

   (angles: $3x+10$ and $2x+40$)

   (A) 30 (B) 50 (C) 70
   (D) 90 (E) 110

2. Two angles are supplementary. If one angle measures x and the other angle measures $x+30$, what is the value of x?

   (A) 60 (B) 65 (C) 70
   (D) 75 (E) 80

3. Two angles are complementary. If the measure of one angle is twice the measure of the other angle, what is the measure of the larger angle?

   (A) 20 (B) 30 (C) 40
   (D) 50 (E) 60

4. Two angles are supplementary. If the ratio of the measure of the smaller angle to that of the larger angle is 5 : 7, what is the measure of the smaller angle?

   (A) 60 (B) 65 (C) 70
   (D) 75 (E) 80

5. Two angles are complementary angles. What is the mean of the complementary angles?

   (A) 30 (B) 45 (C) 60
   (D) 75 (E) 90

SOLOMON ACADEMY — LESSON 10

6. If the two lines are parallel in the figure below, what is the value of $x + y$?

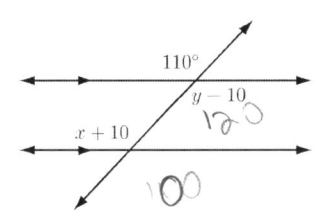

   (A) 220 (B) 200 (C) 180
   (D) 160 (E) 140

7. $\angle A$ and $\angle B$ are complementary angles. $\angle B$ and $\angle C$ are complementary angles. If $m\angle A = 40$, what is the $m\angle C$?

   (A) 30 (B) 35 (C) 40
   (D) 45 (E) 50

8. A straight angle is divided into three smaller angles. If the measures of three smaller angles are consecutive even integers, what is the measure of the largest angle?

   (A) 58 (B) 59 (C) 60
   (D) 61 (E) 62

9. If the two lines are parallel in the figure below, what is the value of x?

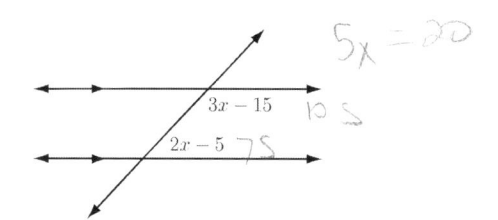

   (A) 40 (B) 45 (C) 50
   (D) 55 (E) 60

10. Angles A and B are supplementary angles. Angles B and C are complementary angles. If the measure of angle B is 10 less than three times the measure of angle C, what is the sum of the measures of angle A and C?

    (A) 150 (B) 140 (C) 130
    (D) 120 (E) 110

ANSWERS AND SOLUTIONS

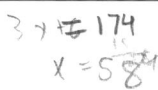

1. (A)

   Vertical angles are congruent. Thus, set $3x + 10$ and $2x + 40$ equal to each other and solve for x.

   $$3x + 10 = 2x + 40$$
   $$x = 30$$

   Therefore, the value of x is 30.

2. (D)

   Since the two angles are supplementary, the sum of their measures is 180°.

   $$x + x + 30 = 180$$
   $$2x + 30 = 180$$
   $$x = 75$$

   Therefore, the value of x is 75.

47

www.solomonacademy.net

3. (E)

Define x as the measure of the smaller angle. Then, $2x$ is the measure of the larger angle. Since the two angles are complementary, the sum of their measures is 90°.

$$x + 2x = 90$$
$$3x = 90$$
$$x = 30$$

Therefore, the measure of the larger angle is $2x = 2(30) = 60°$.

4. (D)

The ratio of the measures of the two angles is 5 : 7. So, let $5x$ be the measure of the smaller angle and $7x$ be the measure of the larger angle. Since the two angles are supplementary, the sum of their measures is 180°.

$$5x + 7x = 180$$
$$12x = 180$$
$$x = 15$$

Therefore, the measure of the smaller angle is $5x = 5(15) = 75°$.

5. (B)

The definition of the mean is the sum of the two numbers divided by 2. Since the two angles are complementary, the sum of their measures is 90°. Therefore, the mean of two complementary angles is $\frac{90}{2} = 45°$.

6. (A)

The angles 110° and $y - 10$ are vertical angles and congruent. Thus,

$$y - 10 = 110$$
$$y = 120$$

Additionally, angles 110° and $x + 10$ are corresponding angles and congruent. Thus,

$$x + 10 = 110$$
$$x = 100$$

Thus, $x = 100$ and $y = 120$. Therefore, the value of $x + y = 220$.

7. (C)

The angles A and B are complementary. Thus, the sum of their measures is 90°. Since the measure of angle A is 40°, the measure of angle B is 50°. Additionally, since angles B and C are complementary and the measure of angle B is 50°, the measure of angle C is 40°. Furthermore, the measure of angle C can be obtained according to the congruent complements theorem.

$$m\angle A + m\angle B = 90°$$
$$m\angle C + m\angle B = 90°$$
$$\therefore \ m\angle A = m\angle C = 40°$$

8. (E)

Let x be the measure of the middle angle. Since the measures of the three angles are consecutive even integer, $x+2$ is the measure of the largest angle and $x-2$ is the measure of the smallest angle. Because the three angles are formed from the straight angle, the sum of their measures is $180°$.

$$x - 2 + x + x + 2 = 180$$
$$3x = 180$$
$$x = 60$$

Therefore, the measure of the largest angle is $x + 2 = 60 + 2 = 62°$.

9. (A)

Since angles $3x - 15$ and $2x - 5$ are consecutive angles, they are supplementary angles whose sum of their measures is $180°$.

$$3x - 15 + 2x - 5 = 180$$
$$5x - 20 = 180$$
$$x = 40$$

Therefore, the value of x is 40.

10. (B)

Angles B and C are complementary angles whose sum of their measures is $90°$. Let x be the measure of angle C. Then, $3x - 10$ is the measure of angle B.

$$3x - 10 + x = 90$$
$$4x - 10 = 90$$
$$x = 25$$

Thus, the measure of angle C is $25°$ and the measure of angle B is $3x - 10 = 3(25) - 10 = 65°$. Since angle A and B are supplementary, the measure of angle A is $180 - 65 = 115°$. Therefore, the sum of the measures of angle A and C is $115 + 25 = 140°$.

LESSON 11

Properties and Theorems of Triangles

A triangle is a figure formed by three segments joining three points called **vertices**. A triangle ABC is expressed as $\triangle ABC$. A triangle can be classified according to its sides or its angles.

Classification by Sides

- Equilateral triangle: All sides are equal in length. The measure of each angle is 60°.

- Isosceles triangle: Two sides are equal in length. If two sides of a triangle are congruent, then the angles (**base angles**) opposite them are congruent as shown in the figure below.

- Scalene triangle: All sides are unequal in length.

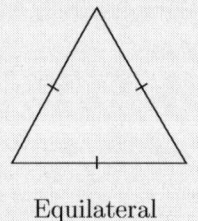

Equilateral   Isosceles   Scalene

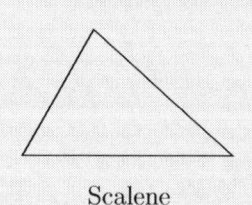

Classification by Angles

- Acute triangle: All interior angles measure less than 90°.

- Right triangle: One of the interior angles measures 90°.

- Obtuse triangle: One of the interior angles measures more than 90°.

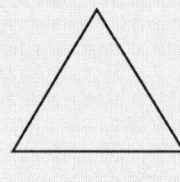

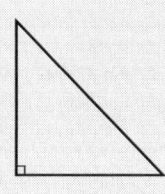

Acute   Right   Obtuse

Area of a Triangle

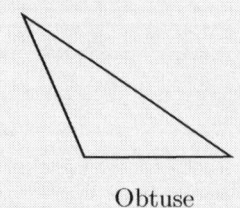

- The area of a triangle is $A = \frac{1}{2}bh$, where b is base and h is height.

- The area of an equilateral triangle with side length of s is $A = \frac{\sqrt{3}}{4}s^2$.

- The areas of two triangles are equal if the bases and heights of the two triangles are the same.

Theorems of triangles

Triangle sum theorem

The sum of the measures of interior angles of a triangle is 180°.

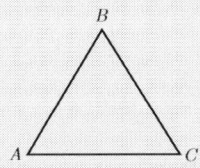

$$m\angle A + m\angle B + m\angle C = 180°$$

Right triangles

In the right triangle, shown at the right, the longest side opposite the right angle is called the **hypotenuse** and the other two sides are called **legs** of the triangle. There is a special relationship between the length of the hypotenuse and the lengths of the legs. It is known as the Pythagorean theorem.

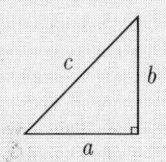

Pythagorean theorem

In the right triangle above, the square of the length of the hypotenuse is equal to the sum of the squares of the lengths of the legs.

$$c^2 = a^2 + b^2$$

The Pythagorean theorem is very useful because it helps you find the length of the third side of a right triangle when the lengths of two sides of the right triangle are known.

45°-45°-90° special right triangles

In a 45°-45°-90° right triangle, the sides of the triangle are in the ratio $1 : 1 : \sqrt{2}$, respectively. In other words, the length of the hypotenuse is $\sqrt{2}$ times the length of each leg.

$$\text{Hypotenuse} = \text{Leg} \times \sqrt{2} \iff \text{Leg} = \frac{\text{Hypotenuse}}{\sqrt{2}}$$

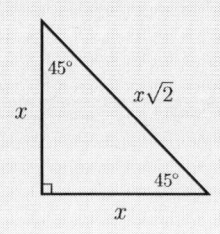

Example: If the measures of the angles of a triangle are $x + 5$, $2x + 10$, and $3x + 15$, what is the measure of the largest angle?

The sum of the measures of interior angles of a triangle is 180°. Thus,

$$x + 5 + 2x + 10 + 3x + 15 = 180$$
$$6x + 30 = 180$$
$$x = 25$$

Therefore, the measure of the largest angle is $3x + 15 = 3(25) + 15 = 90°$.

SOLOMON ACADEMY — LESSON 11

EXERCISES

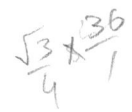

1. What is the area of the equilateral triangle with side length of 6 ?

   (A) $6\sqrt{3}$ (B) $9\sqrt{3}$ (C) $12\sqrt{3}$
   (D) $15\sqrt{3}$ (E) $18\sqrt{3}$

2. In a right triangle, the length of the hypotenuse is 13 and the length of one leg is 5. What is the length of the other leg?

   (A) 4 (B) 6 (C) 8
   (D) 10 (E) 12

3. If the equal sides of an isosceles triangle are $3x+3$ and $2x+7$, what is the length of the equal sides of the isosceles triangle?

   (A) 18 (B) 15 (C) 12
   (D) 10 (E) 8

4. If the area of a square is 100, what is the length of the diagonal?

   (A) $5\sqrt{2}$ (B) $10\sqrt{2}$ (C) $10\sqrt{3}$
   (D) 20 (E) $20\sqrt{2}$

5. In an isosceles triangle, the ratio of the measure of the base angle to that of the vertex angle is 2 : 1, what is the measure of the vertex angle?

   (A) 18 (B) 24 (C) 36
   (D) 48 (E) 72

6. Two trains leave a station at the same time. One train is traveling North at a rate of 30 mph and the other train is traveling East at a rate of 40 mph. After 5 hours, how far apart are they in miles?

   (A) 200 (B) 225 (C) 250
   (D) 300 (E) 350

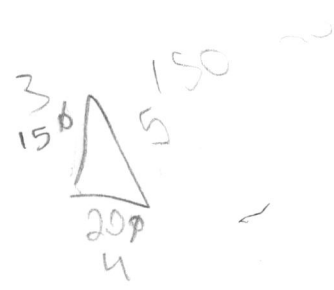

7. If the x and y coordinates of three vertices of a triangle are $(4,0)$, $(9,0)$ and $(2,4)$, what is the area of the triangle?

   (A) 30 (B) 20 (C) 15
   (D) 10 (E) 5

8. In the triangle below, $AC = BC$. What is $m\angle BCD$?

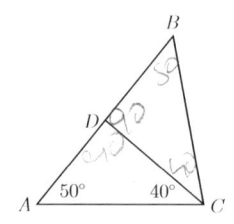

   (A) 40 (B) 45 (C) 50
   (D) 55 (E) 60

9. In an isosceles right triangle ABC, B is the right angle. If $AC = 20$, what is the perimeter of the triangle?

   (A) 20 (B) 40
   (C) $20 + 20\sqrt{2}$ (D) $20 + 30\sqrt{2}$
   (E) $40 + 20\sqrt{2}$

10. In the figure below, $AB = 1$, $BC = 2$, and $CD = 2$. What is AD ?

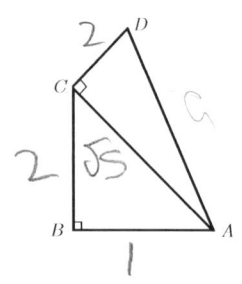

    (A) $\sqrt{29}$ (B) 4 (C) $\sqrt{13}$
    (D) 3 (E) 2

SOLOMON ACADEMY

Distribution or replication of any part of this page is prohibited.

LESSON 11

ANSWERS AND SOLUTIONS

1. (B)

   The area of the equilateral triangle with side length of s is $\frac{\sqrt{3}}{4}s^2$.

   $$\text{Area of equilateral triangle} = \frac{\sqrt{3}}{4}(6)^2 = 9\sqrt{3}$$

   Therefore, the area of the equilateral triangle with side length of 6 is $9\sqrt{3}$.

2. (E)

   Since the triangle is a right triangle, use the Pythagorean theorem: $c^2 = a^2 + b^2$, where c is the hypotenuse, and a and b are the legs of the triangle. The length of the hypotenuse is 13 and the length of one leg is 5. Thus, $c = 13$ and $a = 5$.

$c^2 = a^2 + b^2$	Substitute 13 for c and 5 for a
$13^2 = 5^2 + b^2$	Subtract 25 from each side
$b^2 = 144$	Solve for b
$b = 12$	Since $b > 0$

   Therefore, the length of the other leg is $b = 12$.

3. (B)

   Since $3x + 3$ and $2x + 7$ are the equal sides of the isosceles triangle, set $3x + 3$ and $2x + 7$ equal to each other and solve for x.

   $$3x + 3 = 2x + 7$$
   $$3x - 2x = 7 - 3$$
   $$x = 4$$

   Therefore, the length of the equal sides of the isosceles triangle is $3x + 3 = 3(4) + 3 = 15$.

4. (B)

   In the figure below, the area of the square is 100 which means that the length of the side of the square is 10. The square consists of two 45°-45°-90° triangles whose sides are in the ratio $1 : 1 : \sqrt{2}$.

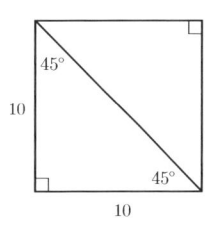

   The diagonal of the square is the hypotenuse of the two triangles. The length of the hypotenuse is $\sqrt{2}$ times the length of each leg. Therefore, the length of the diagonal of the square is $10\sqrt{2}$.

5. (C)

Since the ratio of the measure of the base angle to that of the vertex angle is 2 : 1, let x be the measure of the vertex angle and $2x$ be the measure of the each base angle in the isosceles triangle. Since the sum of the measures of interior angles of triangle BDC is $180°$,

$$x + 2x + 2x = 180$$
$$5x = 180$$
$$x = 36$$

Therefore, the measure of the vertex angle is $36°$.

6. (C)

After 5 hours, the train heading North travels $5 \times 30 = 150$ miles and the other train heading East travels $5 \times 40 = 200$ miles. In order to find out how far they are apart in 5 hours, Use the Pythagorean theorem: $c^2 = 150^2 + 200^2$. Thus, $c = 250$. Therefore, the two trains are 250 miles apart in 5 hours.

7. (D)

In the figure below, the two points $(4, 0)$ and $(9, 0)$ are on the x axis.

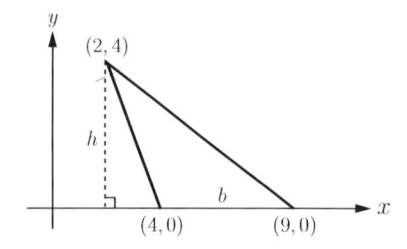

Thus, the base of the triangle is $b = 9 - 4 = 5$. Since the point $(2, 4)$ is 4 units above the x axis, the height of the triangle is $h = 4$. Therefore, the area of the triangle is $A = \frac{1}{2}bh = \frac{1}{2}(5)(4) = 10$.

8. (A)

Since $AC = BC$, triangle ACB is an isosceles triangle. $\angle B$ and $\angle A$ are base angles of the isosceles triangle and are congruent. Thus, $m\angle B = m\angle A = 50°$. Additionally, $\angle BDC$ is an exterior angle of the triangle ADC. Thus, $m\angle BDC = m\angle DAC + m\angle DCA = 90°$. Since the sum of the measures of interior angles of triangle BDC is $180°$,

$$m\angle BCD + m\angle BDC + m\angle B = 180° \quad \text{Substitute } m\angle BDC = 90° \text{ and } m\angle B = 50°$$
$$m\angle BCD + 90° + 50° = 180°$$
$$m\angle BCD = 40°$$

9. (C)

An isosceles right triangle is a $45°$-$45°$-$90°$ right triangle whose sides are in the ratio $1 : 1 : \sqrt{2}$. AC is the length of the hypotenuse of the isosceles right triangle and $AC = 20$. The length of each leg is $\frac{\text{hypotenuse}}{\sqrt{2}} = \frac{20}{\sqrt{2}} = 10\sqrt{2}$. Therefore, the perimeter of the isosceles right triangle is

$$\text{Perimeter of isosceles right triangle} = 20 + 10\sqrt{2} + 10\sqrt{2} = 20 + 20\sqrt{2}$$

10. (D)

In the figure below, find AC using the Pythagorean theorem.

$$AC^2 = AB^2 + BC^2 = 1^2 + 2^2$$
$$AC^2 = 5 \implies AC = \sqrt{5}$$

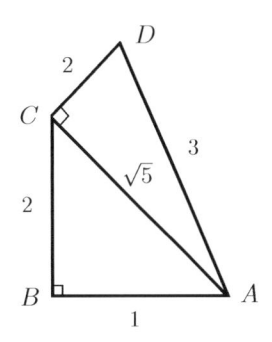

Since AC and CD are known, use the Pythagorean theorem again to find AD.

$$AD^2 = AC^2 + CD^2 = (\sqrt{5})^2 + 2^2$$
$$AD^2 = 9 \implies AD = 3$$

LESSON 12

Patterns and Data Analysis

Patterns

A **pattern** is a set of numbers or objects that are closely related by a specific rule. Understanding a pattern is very important because it helps you predict what will happen next in the set of numbers or objects.

$$\bigcirc \ \triangle \ \square \ \bigcirc \ \triangle \ \square$$

For instance, the figure above shows a pattern which consists of a circle, a triangle, and a square in that order. If the pattern is repeated continuously, how do you predict which one of the three figures is the 36th figure? Since 36 is a multiple of 3 and every third figure is a square, the 36th figure is a square.

Arithmetic sequence and Geometric sequence
There are two most common number patterns: arithmetic sequence and geometric sequence.
A **sequence or progression** is an ordered list of numbers.

- In an arithmetic sequence, add or subtract the same number (common difference) to one term to get the next term.

- In geometric sequence, multiply or divide one term by the same number (common ratio) to get the next term.

- In both sequences, the first term, the second term, and n^{th} term are expressed as a_1, a_2, and a_n respectively.

Type	Definition	Example	n^{th} term
Arithmetic sequence	The common difference between any consecutive terms is constant.	$1, 3, 5, 7, \ldots$	$a_n = a_1 + (n-1)d$ where d is the common difference.
Geometric sequence	The common ratio between any consecutive terms is constant	$2, 4, 8, 16, \ldots$	$a_n = a_1 \times r^{n-1}$ where r is the common ratio.

SOLOMON ACADEMY — LESSON 12

Data Analysis

Mean, or **Average**, is the sum of all elements in a set divided by the number of elements in the set. For instance, if there are 3, 7, and 11 in a set, the mean $= \frac{3+7+11}{3} = 7$.

Median is the middle number when a set of numbers is arranged from least to greatest.

- If there is a n (odd number) number of numbers in a set, the median is the middle number which is $(\frac{n+1}{2})^{th}$ number in the set. For instance, if there are 3, 2, 5, 7, and 10 in a set, arrange the numbers in the set from least to greatest: 2, 3, 5, 7, and 10. Since there are 5 numbers in the set, the median is the $\frac{5+1}{2} = 3^{rd}$ number in the set. Thus, the median is 5.

- If there is a n (even number) number of numbers in a set, the median is the average of the two middle numbers which are the $(\frac{n}{2})^{th}$ and $(\frac{n}{2}+1)^{th}$ numbers. For instance, if there are 1, 4, 6, 8, 9, and 11 in a set, the median is the average of 3rd number and 4th number in the set. Thus, the median is $\frac{6+8}{2} = 7$.

Mode is a number that appears most frequently in a set. It is possible to have more than one mode or no mode in a set.

Range is the difference between the greatest number and the least number in a set.

Example: There are 2, 4, 8, 10, and x in a set. If the range of the set is 13, what is the value of x?

The range is the difference between the greatest number and the least number. The least number in the set is 2. If 10 is the greatest number in the set, the range would be 8. Thus, x must be the greatest number. Therefore, the value of x is 15.

EXERCISES

1. If the pattern below is repeated continuously, what is the 48th letter in the pattern?

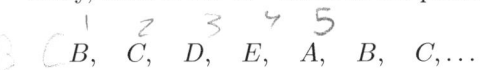

   $A, B, C, D, E, A, B, C, \ldots$

   (A) A (B) B (C) C
   (D) D (E) E

2. In the sequence below, the nth term is defined as $a_n = n^2 + 1$. What is the 7th term in the sequence?

   $2, 5, 10, 17, \ldots$

   (A) 50   (B) 42   (C) 35
   (D) 26   (E) 20

3. In a set of 5, 1, 10, x, and 16, where $10 < x < 16$, which of the following is the median of the set?

   (A) 1   (B) 5   (C) 10
   (D) x (E) 16

4. Jason scored 91, 95, and 94 in the first three tests. What score does he get on the 4th test so that his overall average for the four tests is 94?

   (A) 96   (B) 97   (C) 98
   (D) 99   (E) 100

5. If the volume of a balloon is doubled every three minutes, in how many minutes is the volume of the balloon eight times larger than the initial volume of the balloon?

   (A) 6   (B) 8   (C) 9
   (D) 12   (E) 16

6. In the arithmetic sequence $3, 7, 11, 15, \ldots$ what is the value of the 17th term?

   (A) 76   (B) 67   (C) 56
   (D) 45   (E) 37

7. There are 1, 3, 4, 7, and 10 in a set. If 3 is added to each number in the set, what is the positive difference of the new mean and the new median of the set?

   (A) 1   (B) 2   (C) 3
   (D) 4   (E) 5

8. In a set of five positive integers, the mode is 4, the median is 5, and the mean is 6. What is the greatest of these integers?

   (A) 7   (B) 8   (C) 9
   (D) 10   (E) 11

9. A car travels at 60 miles per hour for 2 hours on a trip and travels at 40 miles per hour for three hours on the returning trip. What is the average speed of the entire trip?

   (A) 55   (B) 50   (C) 48
   (D) 45   (E) 42

10. In the arithmetic sequence, the 3rd term is 17 and the 10th term is 73. What is the 15th term?

    (A) 95   (B) 97   (C) 103
    (D) 108   (E) 113

ANSWERS AND SOLUTIONS

1. (D)

   The pattern consists of B, C, D, E and A. It is repeated continuously. Since every 5th letter is A, the 45th letter is also A. Thus, the 46th letter is B, the 47th letter is C, and the 48th letter is D.

2. (A)

   To find the 7th term in the sequence, substitute 7 for n in $a_n = n^2 + 1$.

   $$a_n = n^2 + 1 \qquad \text{Substitute 7 for } n$$
   $$a_7 = (7)^2 + 1 = 50$$

   Therefore, the 7th term in the sequence is 50.

SOLOMON ACADEMY — Distribution or replication of any part of this page is prohibited. — **LESSON 12**

3. (C)

   Arrange the numbers in the set from least to greatest: 1, 5, 10, x, 16. The median is the middle number of the set. Therefore, the median is 10.

4. (A)

   The average of four tests is 94. This means that Jason should get a total score of $94 \times 4 = 376$ for the four tests. The sum of scores of the first three tests is $91 + 95 + 94 = 280$. Therefore, the score of the fourth test is $376 - 280 = 96$.

5. (C)

   Let's define V_0 as the initial volume of the balloon. The volume of the balloon is doubled every 3 minutes:

   $$\text{In 3 minutes} = 2V_0$$
   $$\text{In 6 minutes} = 2(2V_0) = 4V_0$$
   $$\text{In 9 minutes} = 2(4V_0) = 8V_0$$

   Therefore, in 9 minutes, the volume of the balloon is eight times larger than the initial volume of the balloon.

6. (B)

   In an arithmetic sequence, the first term, $a_1 = 3$ and the common difference, $d = 7 - 3 = 4$. Use the nth term formula to find the 17th term.

   $$a_n = a_1 + (n-1)d \qquad \text{Substitute 17 for } n, 3 \text{ for } a_1, \text{ and 4 for } d$$
   $$a_{17} = 3 + (17-1)4 = 67$$

   Therefore, the value of the 17th term is 67.

7. (A)

   The mean of the set is $\frac{1+3+4+7+10}{5} = 5$. The median of the set is the middle number, 4. If 3 is added to each number in the set, the new mean is $5 + 3 = 8$, and the new median is $4 + 3 = 7$. Therefore, the positive difference of the new mean the new median is $8 - 7 = 1$.

8. (E)

   Since the mean of the five positive integers is 6, the sum of the five positive integers is $5 \times 6 = 30$. Define x as the second greatest integer and y as the greatest integer in the set. Let's consider three cases shown below. In case 1, the median is 4 because there are three 4's. This doesn't satisfy the given information such that the median is 5. Thus, case 1 is false.

Case 1:	$4 + 4 + 4 + x + y = 30$	mode=4, median=4: It doesn't work
Case 2:	$4 + 4 + 5 + 5 + y = 30$	mode=4 and 5, median=5: It doesn't work
Case 3:	$4 + 4 + 5 + x + y = 30$	mode=4, median=5: It works

   In case 2, there are two 4's and two 5's in which the mode are both 4 and 5. This doesn't satisfy the given information such that the mode is 4. Thus, case 2 is false. Finally, in case 3, there are two 4's and one 5. If x is greater than 5, the mode is 4 and median is 5 which satisfy the given information. x must be smallest positive integer greater than 5 so that y will have the greatest possible value. Thus, $x = 6$ and $y = 11$. Therefore, the greatest of these integers is 11.

9. (C)

On the trip, the car travels $60 \times 2 = 120$ miles. On the returning trip, the car travels $40 \times 3 = 120$ miles. Thus,

$$\text{The average speed} = \frac{\text{Total distance}}{\text{Total number of hours}}$$
$$= \frac{120 \text{ miles} + 120 \text{ miles}}{2 \text{ hours} + 3 \text{ hours}}$$
$$= \frac{240 \text{ miles}}{5 \text{ hours}}$$
$$= 48 \text{ miles per hour}$$

Therefore, the average speed of the entire trip is 48 miles per hour.

10. (E)

Write the 10th term and 3rd term of the arithmetic sequence in terms of a_1 and d using the nth term formula: $a_n = a_1 + (n-1)d$.

$$a_{10} = a_1 + 9d = 73$$
$$a_3 = a_1 + 2d = 17$$

Use the linear combinations method to solve for d and a_1.

$$a_1 + 9d = 73$$
$$\underline{a_1 + 2d = 17} \qquad \text{Subtract two equations}$$
$$7d = 56 \qquad \text{Divide both sides by 7}$$
$$d = 8$$

Substitute $d = 8$ in $a_3 = a_1 + 2d = 17$ and solve for a_1. Thus, $a_1 = 1$. Therefore, the 15th term of the arithmetic sequence is $a_{15} = a_1 + 14d = 1 + 14(8) = 113$.

LESSON 13

Counting and Probability

Counting

Counting integers

How many positive integers are there between 42 and 97 inclusive? Are there 54, 55, or 56 integers? Even in this simple counting problem, many students are not sure what the right answer is. A rule for counting integers is as follows:

$$\text{The number of integers} = \text{Greatest integer} - \text{Least integer} + 1$$

According to this rule, the number of integers between 42 and 97 inclusive is $97 - 42 + 1 = 56$ integers.

Counting points on a line

There is a line whose length is 200 feet. If points are placed every 2 feet starting from one end, how many points are on the line?

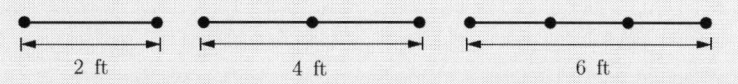

The pattern above suggests that a rule for counting the number of points on a line is as follows:

$$\text{Total number of points on a line} = \frac{\text{Length of a line}}{\text{Distance between each point}} + 1$$

For instance, if a line is 6 feet long, there are $\frac{6}{2} + 1 = 4$ points on the line as shown above. Therefore, if a line is 200 feet long, there are $\frac{200}{2} + 1 = 101$ points on the line.

Counting points on a circle

There is a circle whose circumference is 200 feet. If points are placed every 2 feet on the circumference of the circle, how many points are on the circle?

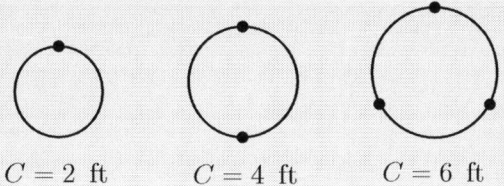

The pattern above suggests that a rule for counting the total number of points on a circle is as follows:

$$\text{Total number of points on a circle} = \frac{\text{Circumference of a circle}}{\text{Distance between each point}}$$

According to this rule, if the circumference of a circle is 200 feet, there are $\frac{200}{2} = 100$ points on the circle.

The fundamental counting principle
If one event can occur in m ways and another event can occur in n ways, then the number of ways both events can occur is $m \times n$. For instance, Jason has three shirts and four pairs of jeans. He can dress up in $3 \times 4 = 12$ different ways.

Venn Diagram
A venn diagram is very useful in counting. It helps you count numbers correctly.

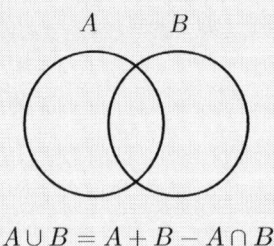

$A \quad B$

$A \cup B = A + B - A \cap B$

In the figure above, $A \cup B$ represents the combined area of two circles A and B. $A \cap B$ represents the common area where the two circles overlap. The venn diagram suggests that the combined area $(A \cup B)$ equals the sum of areas of circles $(A + B)$ minus the common area $(A \cap B)$.

In counting, each circle A and B represents a set of numbers. $n(A)$ and $n(B)$ represent the number of elements in set A and B, respectively. For instance, $A = \{2, 4, 6, 8, 10\}$ and $n(A) = 5$. Thus, the total number of elements that belong to either set A or set B, $n(A \cup B)$, can be counted as follows:

$$n(A \cup B) = n(A) + n(B) - n(A \cap B)$$

Let's find out how many positive integers less than or equal to 20 are divisible by 2 or 3. Define A as the set of numbers divisible by 2 and B as the set of numbers divisible by 3.

$$A = \{2, 4, 6, \cdots, 18, 20\}, \quad n(A) = 10$$
$$B = \{3, 6, 9, 12, 15, 18\}, \quad n(B) = 6$$
$$A \cap B = \{6, 12, 18\}, \quad n(A \cap B) = 3$$

Notice that $A \cap B = \{6, 12, 18\}$ are multiples of 2 and multiples of 3. They are counted twice so they must be excluded in counting. Thus,

$$n(A \cup B) = n(A) + n(B) - n(A \cap B)$$
$$= 10 + 6 - 3$$
$$= 13$$

Therefore, the total number of positive integers less than or equal to 20 that are divisible by 2 or 3 is 13.

SOLOMON ACADEMY — LESSON 13

Probability

The definition of probability of an event, E, is as follows:

$$\text{Probability}(E) = \frac{\text{The number of outcomes event } E \text{ can happen}}{\text{The total number of possible outcomes}}$$

Probability is a measure of how likely an event will happen. Probability can be expressed as a fraction, a decimal, and a percent, and is measured on scale from 0 to 1. Probability can not be less than 0 nor greater than 1.

- Probability equals 0 means an event will never happen.
- Probability equals 1 means an event will always happen.
- Higher the probability, higher chance an event will happen.

For instance, what is the probability of selecting a prime number at random from 1 to 5? In this problem, the event E is selecting a prime number from three possible prime numbers: 2, 3, and 5. The total possible outcomes are numbers from 1 to 5. Thus, the probability of selecting a prime number is $P(E) = \frac{\{2,3,5\}}{\{1,2,3,4,5\}} = \frac{3}{5}$.

Geometric probability

Geometric probability involves the length or area of the geometric figures. The definition of the geometric probability is as follows:

$$\text{Geometric probability} = \frac{\text{Area of desired region}}{\text{Total area}}$$

In the figure below, a circle is inscribed in the square with side length of 10. Assuming that a dart always lands inside the square, what is the probability that a dart lands on a region that lies outside the circle and inside the square?

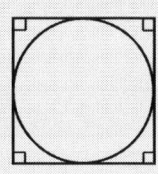

The area of the square is $10^2 = 100$, and the area of the circle is $\pi(5)^2 = 25\pi$. Thus, the area of desired region is $100 - 25\pi$.

$$\begin{aligned}\text{Geometric probability} &= \frac{\text{Area of desired region}}{\text{Total area}} \\ &= \frac{100 - 25\pi}{100} \\ &= \frac{25(4-\pi)}{100} \\ &= \frac{4-\pi}{4}\end{aligned}$$

SOLOMON ACADEMY

Distribution or replication of any part of this page is prohibited.

LESSON 13

Example: You toss a coin four times. How many different outcomes are possible?

Event 1, event 2, event 3, and event 4 are tossing a first coin, second coin, third coin, and fourth coin, respectively. For each event, there are two possible outcomes: head or tail. According to the fundamental counting principle, there are $2 \times 2 \times 2 \times 2 = 16$ possible outcomes for the four events.

EXERCISES

1. There are three red, two yellow, and five blue cards. What is the probability that a blue card is selected at random?

   (A) $\dfrac{3}{5}$ (B) $\dfrac{1}{2}$ (C) $\dfrac{2}{5}$
   (D) $\dfrac{3}{10}$ (E) $\dfrac{1}{5}$

2. How many positive integers are there between 19 and 101 exclusive?

   (A) 78   (B) 79   (C) 80
   (D) 81   (E) 82

3. There are four types of breads, five types of meats, and three types of cheese. Assuming you have to select one of each category, how many different sandwiches can you make?

   (A) 60   (B) 50   (C) 40
   (D) 30   (E) 12

4. Toss a coin twice. What is the probability that you have one head and one tail?

   (A) $\dfrac{1}{4}$ (B) $\dfrac{1}{2}$ (C) $\dfrac{3}{4}$
   (D) $\dfrac{4}{5}$ (E) 1

5. On a road that is 200 feet long, trees are placed every 4 feet starting from one end. How many tree are on the road?

   (A) 49   (B) 50   (C) 51
   (D) 52   (E) 53

6. There are two concentric circles whose radii are 2 and 3, respectively. Assuming a dart never land outside the larger circle, what is the probability that a dart lands on the shaded region?

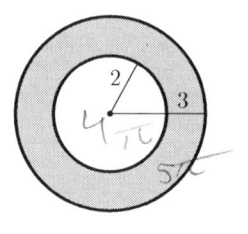

   (A) $\dfrac{1}{3}$ (B) $\dfrac{1}{2}$ (C) $\dfrac{2}{3}$
   (D) $\dfrac{3}{4}$ (E) $\dfrac{5}{9}$

7. There is a square-shaped plot of land whose side length is 50 feet. If posts are placed every 5 feet on the perimeter of the land, how many posts are on the land?

   (A) 39   (B) 40   (C) 41
   (D) 42   (E) 43

8. How many three digit numbers are there whose digits in the hundreds place and ones place are the same? (Assume that a nonzero digit is in the hundreds place.)

   (A) 80   (B) 81   (C) 90
   (D) 100   (E) 121

SOLOMON ACADEMY — Distribution or replication of any part of this page is prohibited. **LESSON 13**

9. There are 36 marbles in a bag. They are either blue, red, or green marbles. The number of green marbles is twice the number of blue marbles. The number of the red marbles is 8 more than the number of blue marbles. What is the probability that a red marble is selected at random in a bag?

   (A) $\frac{1}{3}$ (B) $\frac{13}{36}$ (C) $\frac{7}{18}$
   (D) $\frac{5}{12}$ (E) $\frac{4}{9}$

10. If a number is selected at random from 1 to 30 inclusive, what is the probability that the selected integer is divisible by 2 or 3 ?

    (A) $\frac{8}{15}$ (B) $\frac{9}{15}$ (C) $\frac{2}{3}$
    (D) $\frac{11}{15}$ (E) $\frac{4}{5}$

ANSWERS AND SOLUTIONS

1. (B)

   There are 10 cards. Out of these cards, there are 5 blue cards. Therefore, the probability of selecting a blue card is $\frac{5}{10} = \frac{1}{2}$.

2. (D)

   The two numbers, 19 and 101, are excluded. Therefore, the number of integers from 20 and 100 is $100 - 20 + 1 = 81$.

3. (A)

   Event 1, event 2, and event 3 are selecting one out of 4 types of breads, one out of 5 types of meats, and one out of 3 types of cheese, respectively. According to the fundamental counting principle, you can make $4 \times 5 \times 3 = 60$ different sandwiches.

4. (B)

   Event 1 is tossing a first coin and event 2 is tossing a second coin. According to the fundamental counting principle, there are $2 \times 2 = 4$ outcomes. The four outcomes are $HH, HT, TH,$ and TT. Out of these outcomes, there are two outcomes that have one head and one tail: HT and TH. Therefore, the probability that you have one head and one tail is $\frac{2}{4} = \frac{1}{2}$.

5. (C)

   This problem is exactly the same as counting points on the line. The road is 200 feet long and trees are placed every 4 feet.

   $$\text{Total number of trees on the road} = \frac{\text{Length of the road}}{\text{Distance between each tree}} + 1$$
   $$= \frac{200}{4} + 1$$
   $$= 51$$

   Therefore, there are 51 trees on the road.

65   www.solomonacademy.net

SOLOMON ACADEMY — Distribution or replication of any part of this page is prohibited. — LESSON 13

6. (E)

   The area of the larger circle is $\pi(3)^2 = 9\pi$. The area of the shaded region is $\pi(3)^2 - \pi(2)^2 = 5\pi$.

   $$\text{Geometric probability} = \frac{\text{Area of the shaded region}}{\text{Area of larger circle}} = \frac{5\pi}{9\pi} = \frac{5}{9}$$

   Therefore, the probability that a dart lands on the shaded region is $\frac{5}{9}$.

7. (B)

   This problem is exactly the same as counting points on the circle. The perimeter of the square-shaped plot of land is $50 \times 4 = 200$ feet. The posts are placed every 5 feet.

   $$\text{Total number of posts on the land} = \frac{\text{Perimeter of the land}}{\text{Distance between each post}}$$
   $$= \frac{200}{5}$$
   $$= 40$$

   Therefore, there are 40 posts on the land.

8. (C)

   Event 1 is selecting a digit in the hundreds place and ones place. There are 9 possible outcomes: 1 through 9. Event 2 is a selecting a digit in the tens place. There are 10 possible outcomes: 0 through 9. Therefore, according to the fundamental counting principle, there are $9 \times 10 = 90$ three digit numbers whose digits in hundreds place and the ones place are the same.

9. (D)

   Let x be the number of blue marbles. Then, the number of green marbles is $2x$, and the number of red marble is $x + 8$. Since there are 36 marbles, set up an equation and solve for x.

   $$x + x + 8 + 2x = 36$$
   $$4x + 8 = 36$$
   $$4x = 28$$
   $$x = 7$$

   Thus, the number of the red marble is $x + 8 = 15$. Therefore, the probability that a red marble is selected at random is $\frac{15}{36} = \frac{5}{12}$.

10. (C)

    Let's define A as the set of integers that are divisible by 2, B as the set of integers that are divisible by 3, and $A \cap B$ as the set of integers that are divisible by 2 and 3, respectively.

    $$A = \{2, 4, 6, \cdots, 28, 30\}, \qquad n(A) = 15$$
    $$B = \{3, 6, 9, \cdots, 27, 30\}, \qquad n(B) = 10$$
    $$A \cap B = \{6, 12, 18, 24, 30\}, \qquad n(A \cap B) = 5$$

    Thus, the number of integers less than or equal to 30 that are divisible by 2 or 3, $n(A \cup B)$, is

$$n(A \cup B) = n(A) + n(B) - n(A \cap B)$$
$$= 15 + 10 - 5$$
$$= 20$$

Out of the integers from 1 to 30 inclusive, there are 20 integers that are divisible by 2 or 3. Therefore, the probability that a selected integer is divisible by 2 or 3 is $\frac{20}{30} = \frac{2}{3}$.

SOLOMON ACADEMY   Distribution or replication of any part of this page is prohibited.   PRACTICE TEST 1

PRACTICE TEST 1
MATHEMATICS PROBLEMS
50 Questions
Time — 60 minutes

Directions: Solve each problem and enter your answer by marking the circle on the answer sheet. Choose the best answer among the answer choices given.

1. $25.6 \div 0.4 =$
   A. 0.064
   B. 0.64
   C. 6.4
   D. 64
   E. 640

2. If Joshua scored 93, 89, 96, and 98 on four math tests, what is his mean score for the math tests?
   F. 93
   G. 94
   H. 95
   J. 96
   K. 97

3. How many factors does 120 have?
   A. 8
   B. 10
   C. 12
   D. 14
   E. 16

4. Which of the following expression is equal to $2x(x-3)$?
   F. $3x - 6$
   G. $3x - 5$
   H. $2x^2 - 6x$
   J. $2x^2 - 5x$
   K. $3x^2 - 6x$

5. Between which two consecutive positive integers is $\sqrt{200}$?
   A. 13 and 14
   B. 14 and 15
   C. 15 and 16
   D. 16 and 17
   E. 17 and 18

6. If $2x + 3y = 6$, what is y in terms of x ?
   F. $\frac{2}{3}x - 2$
   G. $\frac{3}{2}x - 2$
   H. $-2x + 6$
   J. $-\frac{3}{2}x + 2$
   K. $-\frac{2}{3}x + 2$

68   www.solomonacademy.net

SOLOMON ACADEMY — Distribution or replication of any part of this page is prohibited. — PRACTICE TEST 1

7. How much greater than 1264 is the value of 1264 rounded to the nearest hundreds?
   - A. 34
   - B. 35
   - C. 36
   - D. 48
   - E. 64

8. If $A = \{10, 2, 8, 6, 12, 4\}$, what is the median of the six elements in set A?
   - F. 6
   - G. 7
   - H. 8
   - J. 9
   - K. 10

9. If $y = -x + 3$, what is $-2y + 4$ in terms of x?
   - A. $2x - 2$
   - B. $2x - 6$
   - C. $3x - 2$
   - D. $x^2 - 2$
   - E. $x^2 + 13$

10. What is the value of $x^2 - y^2$ when $x = 100$ and $y = 99$?
    - F. 199
    - G. 200
    - H. 201
    - J. 202
    - K. 203

11. If $3(x - 2) = 2(x - 2)$, what is the value of x?
    - A. 0
    - B. 1
    - C. 2
    - D. 3
    - E. 4

12. Which of the following expression is equal to $(2x - y)^2$?
    - F. $2x^2 + y^2$
    - G. $2x^2 - y^2$
    - H. $4x^2 + y^2$
    - J. $4x^2 + 4xy + y^2$
    - K. $4x^2 - 4xy + y^2$

13. Which of the following number has only two factors?
    - A. 21
    - B. 27
    - C. 39
    - D. 51
    - E. 53

14. If $y = 3x - 4$ and $3x + y + 4 = 12$, what is the value of x?
    - F. 2
    - G. 3
    - H. 4
    - J. 5
    - K. 6

www.solomonacademy.net

15. If $x - 3y = 0$, what is the value of $\frac{x}{y}$?

    A. $\frac{1}{3}$
    B. $\frac{1}{2}$
    C. 2
    D. 3
    E. 4

16. Which of the following number has the smallest remainder when the number is divided by 7?

    F. 41
    G. 44
    H. 45
    J. 46
    K. 47

17. If the length of a cube is 2, what is the volume of the cube?

    A. 2
    B. 6
    C. 8
    D. 12
    E. 16

18. The sum of x and y is 9. The difference of y and x is 3. If $y < x$, which of the following systems of linear equations can be solved to find the value of x and y?

    F. $x + y = 9$
       $y - x = 3$
    G. $x + y = 9$
       $x - y = 3$
    H. $x + y = 9$
       $xy = 3$
    J. $xy = 9$
       $y - x = 3$
    K. $xy = 9$
       $x - y = 3$

19. How many even numbers are there between 1 and 50 inclusive?

    A. 22
    B. 23
    C. 24
    D. 25
    E. 26

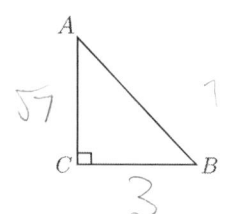

20. In the right triangle above, $AC = \sqrt{7}$ and $CB = 3$. What is the length of $\overline{AB}$?

    F. 4
    G. 5
    H. 6
    J. 7
    K. 8

21. If $2x + 3y = 12$ and $y + 3 = 5$, what is the value of x?

    A. 3
    B. 4
    C. 5
    D. 6
    E. 7

22. What is the slope of a line that passes through the points $(-2, -3)$ and $(2, 9)$?

    F. -3
    G. -2
    H. 1
    J. 2
    K. 3

23. x years ago from now, Joshua was y years old. How old will he be in y years from now?

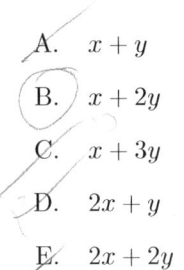

    A. $x + y$
    B. $x + 2y$
    C. $x + 3y$
    D. $2x + y$
    E. $2x + 2y$

x	-1	0	2	5
y	9	4	0	9

24. The table above shows four ordered pairs on the graph of $y = (x - k)^2$. What is the value of k?

    F. -3
    G. -2
    H. 1
    J. 2
    K. 3

25. Joshua walks x feet in y seconds. How many feet will Joshua walk in z minutes?

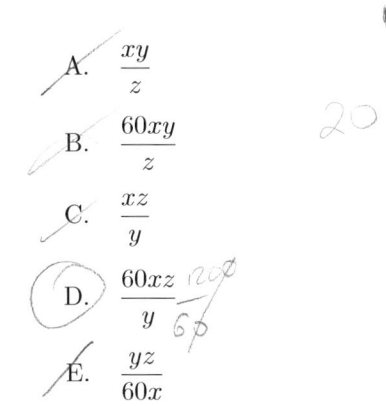

    A. $\dfrac{xy}{z}$
    B. $\dfrac{60xy}{z}$
    C. $\dfrac{xz}{y}$
    D. $\dfrac{60xz}{y}$
    E. $\dfrac{yz}{60x}$

26. If $x - 2 = -7$, what is the value of $\dfrac{8 - 4x}{4}$?

    F. 7
    G. 6
    H. 5
    J. 4
    K. 3

27. If the length of the diagonal of a square is 10, what is the area of the square?

    A. 10
    B. 25
    C. 50
    D. 75
    E. 100

28. What is the distance between point $A(5, 0)$ and point $B(0, 12)$?

    F. 15
    G. 14
    H. 13
    J. 12
    K. 10

$$P = 2R - 3E$$

29. In the equation above, P represents profit, R represents revenue, and E represents expenses. What is the profit when the expenses is $750 and the revenue is $1250?

    A. 150
    B. 250
    C. 350
    D. 550
    E. 750

30. 75% of a number is 20% of 30. What is the number?

    F. 6
    G. 8
    H. 10
    J. 12
    K. 14

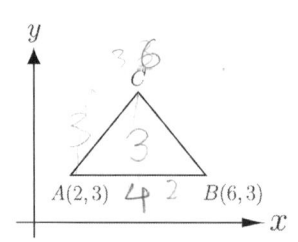

31. In the figure above, $\triangle ABC$ is an isosceles triangle such that $CA = CB$. If the area of $\triangle ABC = 6$, what are the coordinates of point C?

    A. (3, 3)
    B. (3, 6)
    C. (4, 3)
    D. (4, 6)
    E. (4, 9)

$$B = \{9, 6, 11, 5, 16, 12, 15\}$$

32. As shown above, set B has seven positive integers. What is the median of set B?

    F. 12
    G. 11
    H. 10
    J. 9
    K. 8

33. If $9(p-q) = 36$, which of the following value is equal to $6p - 6q$?

    A. 12
    B. 16
    C. 20
    D. 24
    E. 36

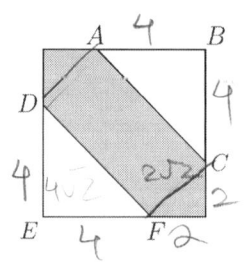

34. In the figure above, points B and E are two vertices of a square with side length of 6. If $\triangle ABC$ and $\triangle DEF$ are isosceles triangles and $AB = DE = 4$, what is the area of the shaded region?

    F. 12
    G. 16
    H. 20
    J. 24
    K. 28

35. The numbers 1 through 10 inclusive are in a hat. If a number is selected at random, what is the probability that the number is neither divisible by 3 nor 4?

   A. $\frac{1}{4}$
   B. $\frac{1}{3}$
   C. $\frac{1}{2}$
   D. $\frac{2}{3}$
   E. $\frac{3}{10}$

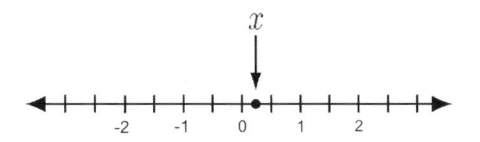

36. As shown above, x is on the number line. Which of the following expression has the largest value?

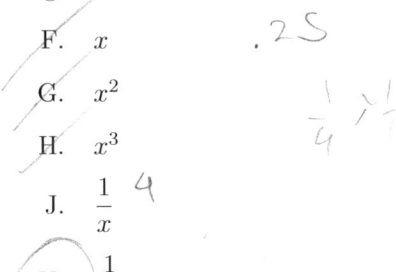

   F. x
   G. x^2
   H. x^3
   J. $\frac{1}{x}$
   K. $\frac{1}{x^2}$

37. If a number, n, is divided by k, the remainder is 5. What is the remainder if $n - k$ is divided by k ?

   A. 1
   B. 2
   C. 3
   D. 4
   E. 5

38. Two numbers are selected at random without replacement from the set $\{1, 2, 3, 4\}$ to form a two-digit number. What is the probability that the two-digit number selected is a prime number?

   F. $\frac{5}{12}$
   G. $\frac{1}{3}$
   H. $\frac{1}{4}$
   J. $\frac{1}{6}$
   K. $\frac{1}{12}$

39. In the sequence above, $3, 7, 11, 15, \ldots$, what is the value of the 13th term?

   A. 50
   B. 51
   C. 52
   D. 53
   E. 54

40. In the figure above, points A, C, and D are on the same line. If $AC : CD = 1 : 2$, what is the ratio of the area of $\triangle ABC$ to that of $\triangle CBD$?

   F. $1 : 9$
   G. $1 : 4$
   H. $1 : 3$
   J. $1 : 2$
   K. $1 : 1$

41. Set S has two numbers and the average of set S is x. Set T has three numbers and the average of set T is y. Which of the following expression represents the average of the two sets, S and T?

A. $\dfrac{x+y}{2}$

B. $\dfrac{x+y}{5}$

C. $\dfrac{xy}{2}$

D. $\dfrac{2x+3y}{2}$

E. $\dfrac{2x+3y}{5}$

42. Let $S(x)$ be the sum of all the positive integers less than or equal to x. For example, $S(5) = 1+2+3+4+5 = 15$. What is the value of $S(15) - S(13)$?

F. 29
G. 35
H. 41
J. 47
K. 52

43. Joshua was born in May. May has 31 days. The statements below describe his birthday. When is Joshua's birthday?

- even number that has 6 factors
- divisible by 3 but not divisible by 9

A. May 30th
B. May 24th
C. May 18th
D. May 12th
E. May 6th

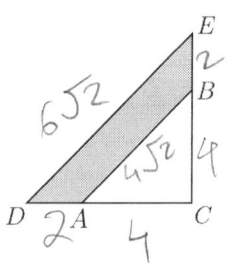

44. In the figure above, $\triangle ABC$ and $\triangle DEC$ are isosceles right triangles. If $AC = 4$ and $DC = 6$, what is the perimeter of the shaded region ABED?

F. $6\sqrt{2}+4$
G. $10\sqrt{2}$
H. $10\sqrt{2}+4$
J. $16\sqrt{2}$
K. $24\sqrt{2}+4$

$$x+y = 5-z$$
$$x-y = 7+z$$

45. In the system of equations above, what is the value of x?

A. 7
B. 6
C. 5
D. 4
E. 3

46. Sue traveled 220 miles at 55 miles per hour. How many minutes longer would the return trip take if she travels at 50 miles per hour?

F. 60 minutes
G. 48 minutes
H. 40 minutes
J. 24 minutes
K. 12 minutes

SOLOMON ACADEMY — PRACTICE TEST 1

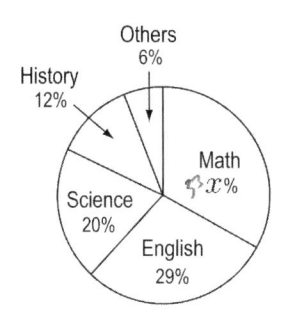

47. As shown above, Joshua made a weekly study plan on the chart that shows the distribution of his study hours over various subjects. If the total study hours is 10 hours in a week, how many hours and minutes will he spend on math?

   A. 2 hours and 18 minutes
   B. 2 hours and 30 minutes
   C. 3 hours and 18 minutes
   D. 3 hours and 30 minutes
   E. 4 hours and 18 minutes

48. ABC represents a three-digit number greater than 200, where $A < B < C$. If B and C are multiples of A, and C is three more than B, which of the following number can be the three-digit number ABC?

   F. 136
   G. 248
   H. 269
   J. 369
   K. 447

49. In square $ABCD$, $\overline{AC}$ and $\overline{BD}$ are the diagonals of the square. What is the product of the slopes of the two diagonals $\overline{AC}$ and $\overline{BD}$?

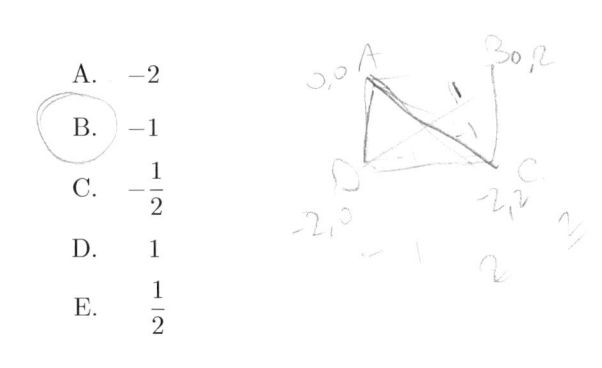

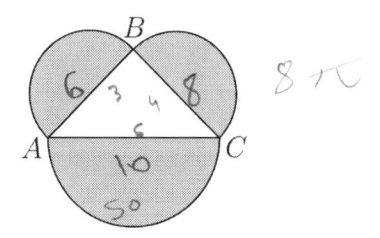

   A. -2
   B. -1
   C. $-\dfrac{1}{2}$
   D. 1
   E. $\dfrac{1}{2}$

Note: Figure not drawn to scale.

50. In $\triangle ABC$ shown above, $AB = 6$, $BC = 8$, and $AC = 10$. There are three semicircles on each side of the triangle. What is the total area of the shaded regions?

   F. 50π
   G. 45π
   H. 35π
   J. 25π
   K. 15π

STOP

SOLOMON ACADEMY — TEST 1 SOLUTIONS

Answers and Solutions
Practice Test 1

Answers

1. D	2. G	3. E	4. H	5. B
6. K	7. C	8. G	9. A	10. F
11. C	12. K	13. E	14. F	15. D
16. G	17. C	18. G	19. D	20. F
21. A	22. K	23. B	24. J	25. D
26. F	27. C	28. H	29. B	30. G
31. D	32. G	33. D	34. H	35. C
36. K	37. E	38. F	39. B	40. J
41. E	42. F	43. D	44. H	45. B
46. J	47. C	48. J	49. B	50. J

Solutions

1. (D)

Multiply the numerator and denominator by 10 and evaluate the expression.

$$\frac{25.6}{0.4} = \frac{25.6 \times 10}{0.4 \times 10} = \frac{256}{4} = 64$$

2. (G)

Tips: Mean, or average, is the sum of all elements in a set divided by the number of elements in the set.

Joshua scored 93, 89, 96, and 98 on four math tests. The sum of the four math scores is $93 + 89 + 96 + 98 = 376$. Therefore, the mean of the four math scores is $\frac{376}{4} = 94$.

3. (E)

Tips: Factors are the numbers that you multiply to get another number. For instance, $15 = 1 \times 15$ and $15 = 3 \times 5$. Thus, the factors of 15 are 1, 3, 5, and 15.

The factors of 120 are 1, 2, 3, 4, 5, 6, 8, 10, 12, 15, 20, 24, 30, 40, 60, and 120. Therefore, there are 16 factors of 120. Or, use the prime factorization of 120. Since $120 = 2^3 \times 3^1 \times 5^1$, the total number of factors is $(3+1) \times (1+1) \times (1+1) = 16$.

4. (H)

Use the distributive property: $a(b - c) = ab - ac$.

$$2x(x - 3) = 2x^2 - 6x$$

SOLOMON ACADEMY

Distribution or replication of any part of this page is prohibited.

TEST 1 SOLUTIONS

5. (B)

> **Tips**
> 1. The larger the number, the greater the value of the square root.
> 2. Use the following perfect squares: $14^2 = 196$ and $15^2 = 225$.

$$196 < 200 < 225 \quad \text{Take the square each side}$$
$$\sqrt{196} < \sqrt{200} < \sqrt{225}$$
$$14 < \sqrt{200} < 15$$

Therefore, $\sqrt{200}$ is between 14 and 15.

6. (K)

$$2x + 3y = 6 \quad \text{Subtract } 2x \text{ from each side}$$
$$3y = -2x + 6 \quad \text{Divide each side by 3}$$
$$y = -\frac{2}{3}x + 2$$

7. (C)

Rounding 1264 to the nearest hundreds, you will get 1300. Therefore, 1300 is 36 greater than 1264.

8. (G)

> **Tips**
> If there is a n (even number) number of numbers in a set, the median is the average of the two middle numbers, which are $\left(\frac{n}{2}\right)^{\text{th}}$ and $\left(\frac{n}{2}+1\right)^{\text{th}}$ numbers. In this problem, the median of the six numbers in set A is the average of the third and fourth numbers.

Rearrange the six numbers in set A from least to greatest: $A = \{2, 4, 6, 8, 10, 12\}$. Since the median is the average of the third and fourth numbers in set A, the median is $\frac{6+8}{2} = 7$.

9. (A)

In order to write $-2y + 4$ in terms of x, substitute $-x + 3$ for y.

$$-2y + 4 = -2(-x + 3) + 4 \quad \text{Substitute } -x + 3 \text{ for } y$$
$$= 2x - 6 + 4$$
$$= 2x - 2$$

Therefore, $-2y + 4$ can be written as $2x - 2$.

10. (F)

> **Tips** Use the difference of squares formula: $x^2 - y^2 = (x+y)(x-y)$

In order to evaluate $x^2 - y^2$ when $x = 100$ and $y = 99$, use the difference of squares formula.

$$x^2 - y^2 = (x+y)(x-y) \quad \text{Substitute 100 for } x \text{ and 99 for } y$$
$$= (100 + 99)(100 - 99)$$
$$= 199$$

11. (C)

$$3(x-2) = 2(x-2)$$ Distribute each side
$$3x - 6 = 2x - 4$$ Subtract $2x$ and add 6 to each side
$$x = 2$$

Therefore, the value of x is 2.

12. (K)

> Tips Use the binomial expansion: $(a-b)^2 = a^2 - 2ab + b^2$

$$(2x-y)^2 = (2x)^2 - 2(2x)(y) + y^2$$
$$= 4x^2 - 4xy + y^2$$

13. (E)

Any prime number has only two factors. Among the answer choices given, 53 is a prime number.

14. (F)

Substitute $3x - 4$ for y in $3x + y + 4 = 12$ and solve for x.

$$3x + y + 4 = 12$$ Substitute $3x - 4$ for y
$$3x + (3x - 4) + 4 = 12$$
$$6x = 12$$
$$x = 2$$

Therefore, the value of x is 2.

15. (D)

$$x - 3y = 0$$ Add $3y$ to each side
$$x = 3y$$ Divide each side by y
$$\frac{x}{y} = 3$$

Therefore, the value of $\frac{x}{y}$ is 3.

16. (G)

The table below shows the remainder when the numbers in the answer choices are divided by 7.

	41	44	45	46	47
Remainder	6	2	3	4	5

Therefore, 44 has the smallest remainder when it is divided by 7.

17. (C)

The volume of the cube with side length of 2 is $2^3 = 8$.

SOLOMON ACADEMY — TEST 1 SOLUTIONS

18. (G)

The sum of x and y is 9 can be expressed as $x + y = 9$. The difference of y and x is 3. Since $y < x$, it can be expressed as $x - y = 3$. Therefore, the system of linear equations in answer choice (G) is the correct answer.

19. (D)

There are 50 integers between 1 and 50 inclusive. Since half of the 50 integers are even integers, there are 25 even integers between 1 and 50.

20. (F)

> **Tips**: If a triangle is a right triangle, use the Pythagorean theorem: $c^2 = a^2 + b^2$, where c is the hypotenuse, a and b are the legs of the right triangle.

Triangle ABC is a right triangle with $CB = 3$ and $AC = \sqrt{7}$. To find the length of the hypotenuse, AB, use the Pythagorean theorem: $AB^2 = 3^2 + (\sqrt{7})^2$. Therefore, the length of $\overline{AB}$ is 4.

21. (A)

Since $y + 3 = 5$, $y = 2$. Substitute 2 for y in the equation $2x + 3y = 12$ and solve for x.

$$2x + 3y = 12 \quad \text{Substitute 2 for } y$$
$$2x + 6 = 12 \quad \text{Solve for } x$$
$$x = 3$$

Therefore, the value of x is 3.

22. (K)

> **Tips**: When two points (x_1, y_1) and (x_2, y_2) are given, the slope $= \frac{y_2 - y_1}{x_2 - x_1}$

Since two points $(-2, -3)$ and $(2, 9)$ are given, use the definition of the slope.

$$\text{Slope} = \frac{y_2 - y_1}{x_2 - x_1} = \frac{9 - (-3)}{2 - (-2)} = 3$$

Therefore, the slope of the line that passes through the points $(-1, 2)$ and $(1, 8)$ is 3.

23. (B)

x years ago from now, Joshua was y years old, which implies that he is $x + y$ years old now. Therefore, in y years from now, Joshua will be $x + y + y$ or $x + 2y$ years old.

24. (J)

To find the value of k, let's select the ordered pair $(2, 0)$.

$$y = (x - k)^2 \quad \text{Substitute 2 for } x \text{ and 0 for } y$$
$$0 = (2 - k)^2 \quad \text{Take the square root of each side}$$
$$2 - k = 0$$
$$k = 2$$

Therefore, the values of k is 2.

SOLOMON ACADEMY — TEST 1 SOLUTIONS

25. (D)

There are 60 seconds in one minute. Thus, there are $60 \times z$ or $60z$ seconds in z minutes. Define n as the number of feet that Joshua will walk in z minutes. Set up a proportion in terms of feet and seconds and solve for n.

$$x_{\text{feet}} : y_{\text{seconds}} = n_{\text{feet}} : 60z_{\text{seconds}}$$

$$\frac{x}{y} = \frac{n}{60z} \qquad \text{Use cross product property}$$

$$ny = 60xz \qquad \text{Solve for } n$$

$$n = \frac{60xz}{y}$$

Therefore, Joshua will walk $\frac{60xz}{y}$ feet in z minutes.

26. (F)

Since $x - 2 = -7$, $x = -5$. Substitute -5 for x in the expression.

$$\frac{8 - 4x}{4} = \frac{8 - 4(-5)}{4} = \frac{28}{4} = 7$$

Therefore, the value of $\frac{8-4x}{4}$ is 7.

27. (C)

> Tips: If d is the length of the diagonal of a square, the area of the square is $\frac{1}{2}d^2$.

Since the length of the diagonal of the square is 10,

$$\text{Area of square} = \frac{1}{2}d^2 = \frac{1}{2}(10)^2 = 50$$

28. (H)

> Tips: Given the two points (x_1, y_1) and (x_2, y_2), the distance formula is as follows:
> $$\text{distance} = \sqrt{(x_2 - x_1)^2 + (y_2 - y_1)^2}$$

Use the distance formula to find the distance between points $A(5,0)$ and $B(0,12)$.

$$d = \sqrt{(x_2 - x_1)^2 + (y_2 - y_1)^2} = \sqrt{(0-5)^2 + (12-0)^2} = 13$$

Therefore, the distance between points A and B is 13.

29. (B)

To find P(profit), plug in the given values of R(revenue) and E(expenses) into the equation $P = 2R - 3E$. Since $R = \$1250$ and $E = \$750$,

$$P = 2R - 3E = 2(1250) - 3(750) = 250$$

Therefore, when the expenses is $750 and the revenue is $1250, the profit of the company is $250.

SOLOMON ACADEMY — TEST 1 SOLUTIONS

30. (G)

1% means 1 out of 100. Thus, 75% is equal to $\frac{75}{100} = \frac{3}{4}$ and 20% is equal to $\frac{20}{100} = \frac{1}{5}$. Let x be the number. 75% of a number is 20% of 30 can be written as $\frac{3}{4}x = \frac{1}{5} \times 30$.

$$\frac{3}{4}x = \frac{1}{5} \times 30 \qquad \text{Solve for } x$$
$$x = 8$$

Therefore, the number is 8.

31. (D)

> **Tips**
> Given the two points (x_1, y_1) and (x_2, y_2), the midpoint formula is as follows:
> $$\text{midpoint} = \left(\frac{x_1 + x_2}{2}, \frac{y_1 + y_2}{2}\right)$$

In the figure below, $\triangle ABC$ is an isosceles triangle such that $CA = CB$. Draw a segment from vertex C to point D such that $\overline{CD} \perp \overline{AB}$. Then, point D is the midpoint between two points $A(2,3)$ and $B(6,3)$.

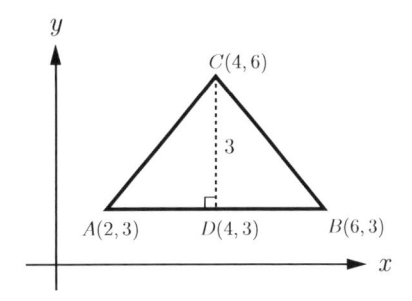

To find the x and y coordinates of the midpoint, D, use the midpoint formula: $\left(\frac{x_1+x_2}{2}, \frac{y_1+y_2}{2}\right)$. Thus, the midpoint, D, is $\left(\frac{2+6}{2}, \frac{3+3}{2}\right)$ or $(4,3)$. The area of $\triangle ABC$ is 6 and the length of the base, AB, is 4. Thus, the height of $\triangle ABC$, CD, must be 3, which implies that point C is 3 units directly above point $D(4,3)$. Therefore, the x and y coordinates of point C is $(4,6)$.

32. (G)

To find the median of set B, arrange the numbers from least to greatest. Thus,

$$B = \{5, 6, 9, 11, 12, 15, 16\}$$

Since there are seven numbers in set B, the median is the middle number which is the fourth number in set B. Therefore, the median of set B is 11.

33. (D)

$$9(p - q) = 36 \qquad \text{Divide each side by 9}$$
$$p - q = 4 \qquad \text{Multiply each side by 6}$$
$$6(p - q) = 24 \qquad \text{Expand}$$
$$6p - 6q = 24$$

34. (H)

In the figure below, △ABC is an isosceles right triangle with $AB = BC = 4$. The area of △ABC is $\frac{1}{2}bh$ or 8. Since $AB = DE = 4$, △ABC and △DEF are congruent. Thus, the area of △DEF is also 8.

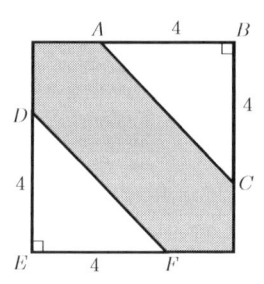

The area of the shaded region equals the area of the square minus the sum of the areas of △ABC and △DEF. Therefore,

$$\text{Area of shaded region} = \text{Area of square} - (\text{Area of } \triangle ABC + \text{Area of } \triangle DEF)$$
$$= 6^2 - (8 + 8)$$
$$= 20$$

35. (C)

There are three numbers that are divisible by 3: 3, 6, and 9. Also, there are two numbers that are divisible by 4: 4, and 8. Thus, there are five numbers that are divisible by three or four. Since there are 10 numbers in the hat, five numbers are neither divisibly by 3 nor 4. Therefore, the probability that the number is neither divisible by 3 nor 4 is $\frac{5}{10}$ or $\frac{1}{2}$.

36. (K)

For simplicity, let's choose $\frac{1}{4}$ for the value of x. Plug in $\frac{1}{4}$ into the value of x for each answer choice to find out which of the following expression has the largest value.

F. $x = \dfrac{1}{4}$

G. $x^2 = \dfrac{1}{16}$

H. $x^3 = \dfrac{1}{64}$

J. $\dfrac{1}{x} = \dfrac{1}{\frac{1}{4}} = 4$

K. $\dfrac{1}{x^2} = \dfrac{1}{\frac{1}{16}} = 16$

Therefore, (K) is the correct answer.

SOLOMON ACADEMY   Distribution or replication of any part of this page is prohibited.   TEST 1 SOLUTIONS

37. (E)

> **Tips**: When n is divided by k, the quotient is q and the remainder is r. Then, n can be written as $n = kq + r$

When 100 is divided by 3, the quotient is 33 and the remainder is 1. Thus, 100 can be expressed as $100 = 3 \times 33 + 1$. Likewise, when n is divided by k, the quotient is q and the remainder is 5. Thus, n can be expressed as $n = kq + 5$. To find the remainder when $n - k$ is divided by k,

$$n = kq + 5 \qquad \text{Subtract } k \text{ from each side}$$
$$n - k = kq - k + 5 \qquad \text{Factor } kq - k$$
$$n - k = k(q - 1) + 5$$

$n - k = k(q - 1) + 5$ implies that the remainder is 5 when $n - k$ is divided by k.

38. (F)

> **Tips**: The fundamental counting principle: If one event can occur in m ways and another event can occur n ways, then the number of ways both events can occur is $m \times n$.

Two numbers are selected at random without replacement from the set $\{1, 2, 3, 4\}$ to form a two-digit number. To find the total number of two-digit numbers, use the fundamental counting principle. Define event 1 and event 2 as selecting a digit for the tens' place, and ones' place, respectively. Event 1 has 4 ways to select a digit out of 4 digits. After one digit is taken, event 2 has 3 ways to select a digit out of the three remaining digits. Thus, there are $4 \times 3 = 12$ possible two-digit numbers using 1, 2, 3, and 4, which are shown below.

12 possible two-digit numbers: 12, 13, 14, 21, 23, 24, 31, 32, 34, 41, 42, and 43

Out of 12 possible two-digit numbers, there are 5 prime numbers: 13, 23, 31, 41, and 43. Therefore, the probability that the two-digit number selected is a prime number is $\frac{5}{12}$.

39. (B)

> **Tips**: In an arithmetic sequence, add or subtract the same number (common difference) to one term to get the next term. The nth term of the arithmetic sequence is as follows: $a_n = a_1 + (n - 1)d$, where a_1 is the first term and d is the common difference.

The sequence $3, 7, 11, 15, \cdots$ is an arithmetic sequence, where the first term, $a_1 = 3$ and the common difference, $d = 7 - 3 = 4$. Use the nth term formula to find the 13th term.

$$a_n = a_1 + (n - 1)d \qquad \text{Substitute 13 for } n$$
$$a_{13} = 3 + (13 - 1)4 = 51$$

Therefore, the value of the 13th term is 51.

SOLOMON ACADEMY Distribution or replication of any part of this page is prohibited. TEST 1 SOLUTIONS

40. (J)

In the figure below, points A, C, and D are on the same line. Since $AC : CD = 1 : 2$, let x be the length of $\overline{AC}$ and $2x$ be the length of $\overline{CD}$.

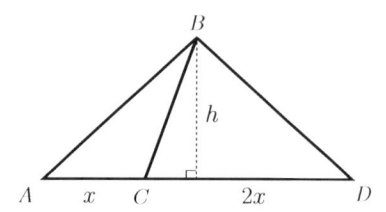

$\triangle ABC$ and $\triangle CBD$ have the same height, h. Thus,

$$\frac{\text{Area of } \triangle ABC}{\text{Area of } \triangle CBD} = \frac{\frac{1}{2} \times AC \times h}{\frac{1}{2} \times CD \times h} \quad \text{Since } AC = x \text{ and } CD = 2x$$

$$= \frac{\frac{1}{2}xh}{\frac{1}{2}(2x)h} \quad \frac{1}{2}xh \text{ cancels out}$$

$$= \frac{1}{2}$$

Therefore, the ratio of the area of $\triangle ABC$ to that of $\triangle CBD$ is $1 : 2$.

41. (E)

(Tips) Sum of elements in a set = average of elements × number of elements

Set S has two numbers and the average of the set is x. Thus, the sum of the two numbers in set S is $2x$. Additionally, Set T has three numbers and the average of the set is y. Thus, the sum of the three numbers in set T is $3y$. The total sum of the numbers in sets S and T is $2x + 3y$. There are five numbers in total in sets S and T. To find the average of sets S and T, divide the total sum of numbers in sets S and T by 5.

$$\text{Average of sets } S \text{ and } T = \frac{\text{Total sum}}{5} = \frac{2x + 3y}{5}$$

Therefore, the average of the two sets, S and T, is $\frac{2x+3y}{5}$.

42. (F)

$S(x)$ is the sum of all positive integers less than or equal to x. Thus,

$$S(15) = 1 + 2 + \cdots + 12 + 13 + 14 + 15$$
$$S(13) = 1 + 2 + \cdots + 12 + 13 \quad \text{Subtract } S(13) \text{ from } S(15)$$
$$\overline{S(15) - S(13) = 14 + 15 = 29}$$

Therefore, the value of $S(15) - S(13)$ is 29.

43. (D)

Use the answer choices to determine which number satisfies all of the given conditions. Eliminate answer choice (C) because 18 is divisible by 9. Eliminate answer choices (A) and (B) because both 30 and 24 have 8 factors. Additionally, eliminate answer choice (E) because 6 has only 4 factors. Therefore, (D) is the correct answer.

44. (H)

In the figure below, $\triangle ABC$ and $\triangle DEC$ are isosceles right triangles. In other words, they are $45°$-$45°$-$90°$ special right triangles whose sides are in the ratio $1 : 1 : \sqrt{2}$.

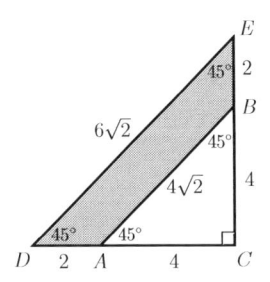

In $\triangle ABC$, $AC = 4$. Thus, $BC = 4$ and $AB = 4\sqrt{2}$. In $\triangle DEC$, $DC = 6$. Thus, $EC = 6$ and $DE = 6\sqrt{2}$. Therefore,

$$\begin{aligned}\text{Perimeter of shaded region} &= DE + EB + AB + DA \\ &= 6\sqrt{2} + 2 + 4\sqrt{2} + 2 \\ &= 10\sqrt{2} + 4\end{aligned}$$

45. (B)

Add the two equations to eliminate the variables y and z. Then, solve for x.

$$\begin{aligned} x + y &= 5 - z \\ x - y &= 7 + z \quad \text{Add two equations} \\ \hline 2x &= 12 \\ x &= 6 \end{aligned}$$

Therefore, the value of x is 6.

46. (J)

> **Tips** Use the time formula: time = $\frac{\text{distance}}{\text{speed}}$

Since $\frac{220 \text{ miles}}{55 \text{ miles per hour}} = 4$ hours, it took Sue 4 hours to travel 220 miles. On the return trip, if she travels at 50 miles per hour, it will take her $\frac{220 \text{ miles}}{50 \text{ miles per hour}} = 4.4$ hours. This means that it will take her 0.4 hour longer on the return trip. Since 0.4 hour is equal to 0.4×60 minutes = 24 minutes, it will take Sue 24 minutes longer on the return trip.

SOLOMON ACADEMY
Distribution or replication of any part of this page is prohibited.

TEST 1 SOLUTIONS

47. (C)

    Joshua plans to spend 29% of his time studying English, 20% on science, 12% on history, and 6% on others. Thus, the percentage of his time studying math can be determined by 100% − (29 + 20 + 12 + 6)% or 33%. Since the total study hours is 10 hours, the total number of hours that Joshua plans to study math is 10 × 0.33 = 3.3 hours. Since 0.3 hour is equal to 0.3 × 60 minutes = 18 minutes, Joshua will spend 3 hours and 18 minutes on math.

48. (J)

    Use the answer choices to determine which number satisfies the given conditions. ABC represents a three-digit number greater than 200. Eliminate answer choice (F) because 136 is less than 200. Digits B and C are multiples of digit A. Thus, eliminate both answer choices (H) and (K) because they do not satisfy it. The remaining answer choices are (G) and (J) Since digit C is three more than digit B, eliminate answer choice (G) because 8 is not three more than 4. Therefore, (J) is the correct answer.

49. (B)

    The diagonals of square $ABCD$, $\overline{AC}$ and $\overline{BD}$, are perpendicular to each other. This means that $\overline{AC}$ and $\overline{BD}$ have negative reciprocal slopes. Thus, the product of the slopes equals −1. For instance, if the slope of $\overline{AC}$ is 1, the slope of $\overline{BD}$ is −1, and the product of the slopes is $1 \times -1 = -1$. Therefore, the product of the slopes of the two diagonals, $\overline{AC}$ and $\overline{BD}$, is −1.

50. (J)

    In $\triangle ABC$ shown below, $AB = 6$, $BC = 8$, and $AC = 10$. There are three semicircles on each side of the triangle. Thus, the radii of the three semicircles are 3, 4, and 5.

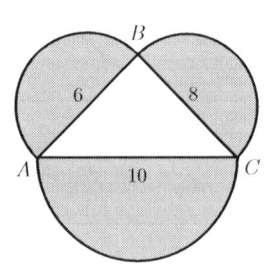

    To find the total area of the shaded regions which equals the sum of the areas of the three semicircles, use the area of a semi-circle formula: $\frac{1}{2}\pi r^2$.

    $$\text{Total area of shaded regions} = \text{Sum of the areas of three semicircles}$$
    $$= \frac{1}{2}\pi(3)^2 + \frac{1}{2}\pi(4)^2 + \frac{1}{2}\pi(5)^2$$
    $$= 25\pi$$

    Therefore, the total area of the shaded regions is 25π.

PRACTICE TEST 2

MATHEMATICS PROBLEMS
50 Questions
Time — 60 minutes

Directions: Solve each problem and enter your answer by marking the circle on the answer sheet. Choose the best answer among the answer choices given.

1. $5\frac{1}{2} \div 2\frac{3}{4} =$

    A. 1
    B. 2
    C. $2\frac{3}{4}$
    D. 3
    E. $3\frac{3}{4}$

2. Express $5 \times 10^3 \times 4.4 \times 10^2$ in scientific notation.

    F. 2.2×10^7
    G. 22×10^6
    H. 22×10^5
    J. 2.2×10^6
    K. 2.2×10^5

3. If the prices of gasoline in the last four months were $3.25, $3.75, $3.50, and $3.90, what is the average price of the gasoline?

    A. $3.50
    B. $3.55
    C. $3.60
    D. $3.65
    E. $3.70

4. What is the value of $\frac{3^3+3^2}{2^2+2}$?

    F. 2
    G. 3
    H. 4
    J. 5
    K. 6

5. If $y = 2x + 4$, what is x in terms of y ?

    A. $y - 4$
    B. $y + 4$
    C. $\frac{y+4}{2}$
    D. $\frac{1}{2}y - 2$
    E. $-\frac{1}{2}y - 2$

6. Set A has five numbers: $t, s, y, z,$ and x. If $x < y < z < s < t$, what is the median of set A ?

    F. x
    G. y
    H. z
    J. t
    K. s

7. If $3(x-2)+5 = 23$, what is the value of x?

   A. 8
   B. 7
   C. 6
   D. 5
   E. 4

8. If $y = 2t$ and $2x - 4y - 12t = 0$, what is x in terms of t?

   F. $4t+6$
   G. $6t-4$
   H. $8t$
   J. $8t+6$
   K. $10t$

9. If $y = (x-1)(x-2)$ and $x = -2$, what is the value of y?

   A. -12
   B. -6
   C. 0
   D. 6
   E. 12

10. Express $10 + \frac{2}{10} + \frac{3}{1000}$ in decimal representation.

    F. 10.23
    G. 10.53
    H. 10.503
    J. 10.203
    K. 10.023

11. If $2x = 3y = 6z = 12$, what is the mean of x, y, and z?

    A. 1
    B. 2
    C. 3
    D. 4
    E. 5

12. What is the sum of all prime numbers less than 10?

    F. 15
    G. 17
    H. 18
    J. 26
    K. 37

13. What is the remainder when 111 is divided by 13?

    A. 5
    B. 7
    C. 8
    D. 9
    E. 10

14. If $\frac{12}{x} + 5 = 9$, what is the value of x?

    F. 3
    G. 4
    H. 6
    J. 8
    K. 12

15. If $2^x = b$, what is 2^{x+3} in term of b?

    A. $b + 3$
    B. $b + 6$
    C. $3b$
    D. $6b$
    E. $8b$

16. If the pattern below is repeated continuously, what is the 27th letter?

    $$F, G, H, J, K, F, G, \cdots$$

    F. F
    G. G
    H. H
    J. J
    K. K

17. Evaluate $(\sqrt{x} - \sqrt{y})^2$ when $x = 16$ and $y = 49$?

    A. 33
    B. 21
    C. 12
    D. 9
    E. 3

18. Joshua is packing for a trip to New York. He has three suitcases that weigh 32 pounds each and two backpacks that weigh 16 pounds each. What is the total weight of his luggage in pounds?

    F. 67
    G. 75
    H. 96
    J. 123
    K. 128

19. If car A is traveling 65 miles per hour, what is the total distance that car A would travel in four hours?

    A. 280 miles
    B. 260 miles
    C. 200 miles
    D. 140 miles
    E. 100 miles

20. If the area of each face of a cube is 16, what is the volume of the cube?

    F. 16
    G. 32
    H. 64
    J. 72
    K. 96

21. A tulip grows 2 inches in a week. At this rate, how many weeks will it take the tulip to grow 2 feet?

    A. 6 weeks
    B. 8 weeks
    C. 10 weeks
    D. 12 weeks
    E. 16 weeks

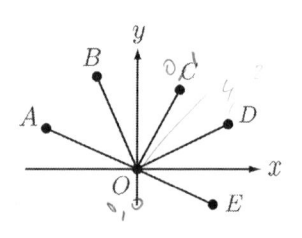

22. In the graphs above, which of the following segment has the largest value of slope?

   F. $\overline{OC}$
   G. $\overline{OD}$
   H. $\overline{OE}$
   J. $\overline{OA}$
   K. $\overline{OB}$

23. How many positive integers less than twenty are prime numbers?

   A. 8
   B. 7
   C. 6
   D. 5
   E. 4

24. If $x = y + 7$, what is the value of $y - x$?

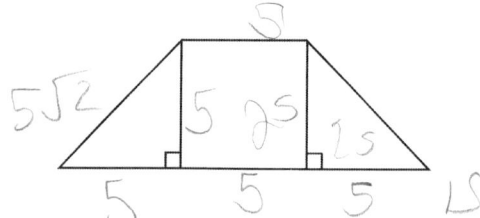

   F. -7
   G. -3
   H. 0
   J. 3
   K. 7

25. In the figure above, the trapezoid consists of two isosceles right triangles and a square. If the length of the hypotenuse of the right triangle is $5\sqrt{2}$, what is the area of the trapezoid?

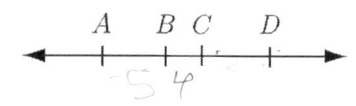

   A. 125
   B. 100
   C. 75
   D. 50
   E. 25

26. If $BC = 4$ and $AD = 9$, what is the value of $AB + CD$?

   F. 1
   G. 2
   H. 3
   J. 4
   K. 5

27. The perimeter of an equilateral triangle is the same as the perimeter of a square with side length of 9. What is the length of the side of the equilateral triangle?

   A. 8
   B. 9
   C. 10
   D. 12
   E. 15

28. Point $A(-1,5)$ and point $B(3,-3)$ are located on the xy-coordinate plane. Where is the midpoint between point A and point B located?

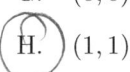

    F. $(-2,4)$
    G. $(0,0)$
    H. $(1,1)$
    J. $(2,2)$
    K. $(4,6)$

29. A clock is malfunctioning. The minute hand of the clock only moves and indicates correct time every 12 minutes. For instance, the clock indicates 12 pm between 12 pm to 12:11 pm and indicates correct time at 12:12 pm. How many times does the clock indicate correct time between 12:10 pm and 5:35 pm?

    A. 23 times
    B. 24 times
    C. 25 times
    D. 26 times
    E. 27 times

30. At store A, two pens cost $1.73. However, at store B, two pens cost $1.99. How much money, in cents, do you save per pen if you buy two pens at store A?

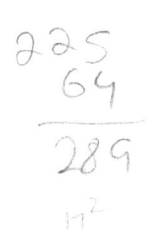

    F. 0.13
    G. 0.26
    H. 0
    J. 13
    K. 26

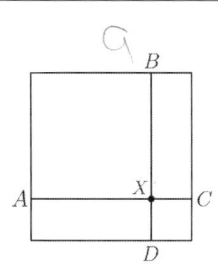

31. In the figure above, point X lies inside a square with side length of 9. If $\overline{AC}$ and $\overline{BD}$ are parallel to the sides of the square, what is the value of $XA + XB + XC + XD$?

    A. 18
    B. 16
    C. 13
    D. 10
    E. 9

32. If $(x+2)^3 = 27$, what is the value of $\dfrac{1}{x+3}$?

    F. $\frac{1}{5}$
    G. $\frac{1}{4}$
    H. $\frac{1}{3}$
    J. 2
    K. 4

33. The Pythagorean theorem states that in a right triangle, the square of the length of the hypotenuse is equal to the sum of the squares of the lengths of the legs. Which of the following is a set of three positive integers that satisfies the Pythagorean theorem?

    A. $1, 1, \sqrt{2}$
    B. $6, 8, 9$
    C. $5, 12, 15$
    D. $7, 23, 25$
    E. $8, 15, 17$

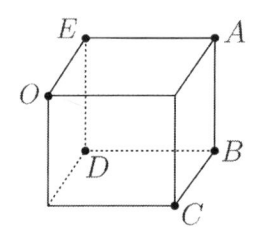

34. In the rectangular box shown above, the length, width, and height of the box are unknown. Which of the following segment is the longest?

F. $\overline{OA}$
G. $\overline{OB}$
H. $\overline{OC}$
J. $\overline{OD}$
K. $\overline{OE}$

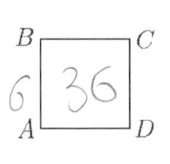

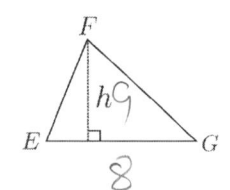

35. In the figure above, square $ABCD$ and triangle EFG have the same areas. If $AB = 6$ and $EG = 8$, what is the height of triangle EFG?

A. 5
B. 6
C. 7
D. 8
E. 9

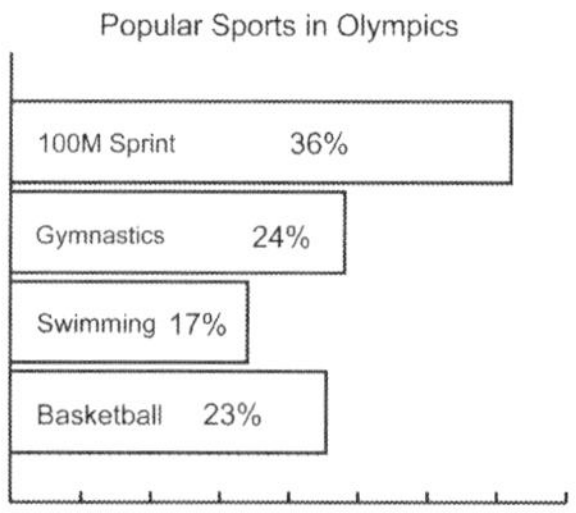

36. A survey asks people in the US which sport they are going to watch in the upcoming Olympics. The chart above shows the results of the survey. If the US population is 300 million, how many more people, in million, are going to watch the most popular sport compared to the least popular sport in the chart?

F. 17
G. 19
H. 21
J. 36
K. 57

37. Mr. Rhee is 5 ft 9 inches tall. His son, Jason, is $\frac{1}{3}$ of Mr. Rhee's height. What is the difference of their heights in inches?

A. 52 inches
B. 46 inches
C. 36 inches
D. 32 inches
E. 23 inches

38. If $y > 0$, the equation $\sqrt{x+y} = \sqrt{x} + \sqrt{y}$ is only true when $x = $?

F. -2
G. -1
H. 0
J. 1
K. 2

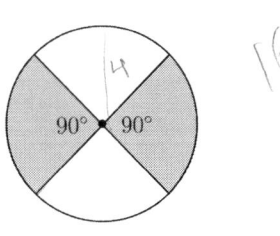

39. In the figure above, the radius of the circle is 4. What is the area of the shaded region?

   A. 4π
   B. 6π
   C. 8π
   D. 10π
   E. 12π

40. What is the sum of the units digit of the first five positive squares?

   F. 17
   G. 18
   H. 20
   J. 22
   K. 25

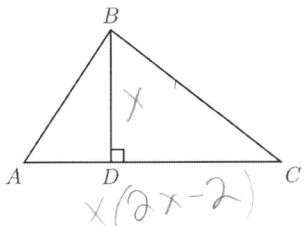

41. In the figure above, $BD = x$ and $AC = 2x - 2$. Which of the following expression best represents the area, A, of $\triangle ABC$?

   A. $A = 3x - 2$
   B. $A = x^2 - 2$
   C. $A = x^2 - x$
   D. $A = 2x^2 - 2$
   E. $A = 2x^2 - 2x$

42. In a set of five distinct positive integers, the average of the two smallest integers is 2, the average of the three smallest integers is 3, the average of the four smallest is 4, and the average of all five integers is 5. What is the largest integer in the set?

   F. 9
   G. 10
   H. 11
   J. 12
   K. 13

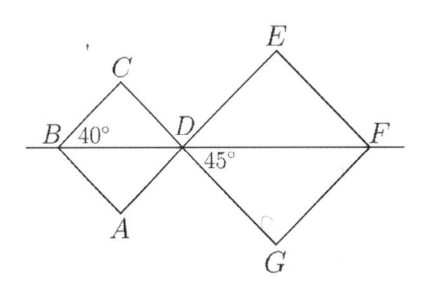

43. In the figure above, two quadrilaterals $ABCD$ and $DEFG$ are parallelograms. What is the measure of $\angle EFG$?

   A. 75
   B. 80
   C. 85
   D. 90
   E. 105

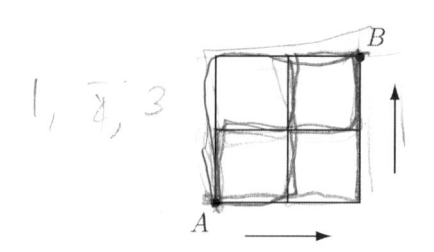

44. As shown in the grid above, Joshua is walking to point B from point A. He is only allowed to walk either right or up, neither left nor down. How many distinct 4-unit paths are there from point A to point B?

F. 3
G. 4
H. 5
J. 6
K. 7

$$x^2 - y^2 = 8$$
$$x - y = 2$$

45. From the equations above, what is the value of $x^2 + y^2$?

A. 25
B. 19
C. 13
D. 10
E. 4

I. The solutions to $x^2 = 81$ are 9 or -9.

II. The values of $\sqrt{4}$ are 2 or -2.

III. The value of $\frac{1}{0}$ is undefined.

IV. If $x < 0$, $|x| = x$.

46. Which of the following statements above are true?

F. I only
G. I & III
H. I & IV
J. II & III
K. I & III & IV

47. Mr. Rhee, Sue, Joshua, and Jason are taking a family photo on a long sofa. What is the probability that Mr. Rhee and Sue sit next to each other?

A. $\frac{1}{6}$
B. $\frac{1}{5}$
C. $\frac{1}{4}$
D. $\frac{1}{3}$
E. $\frac{1}{2}$

48. A big water bottle on the cooler was full on Monday. Students drank one-fourth the water on Tuesday, one-third of the remaining on Wednesday, one-half of the remaining on Thursday. What fractional part of the water would be remaining in the water bottle?

F. $\frac{1}{2}$
G. $\frac{1}{3}$
H. $\frac{1}{4}$
J. $\frac{1}{5}$
K. $\frac{2}{5}$

49. A farmer wants to build a fence around his rectangular field. In addition, he wants to divide it in half with a fence perpendicular to the longer side of the rectangular field so that it becomes two smaller squares. If the total length of the fence is 420 feet, what is the length of the shorter side of the rectangular field?

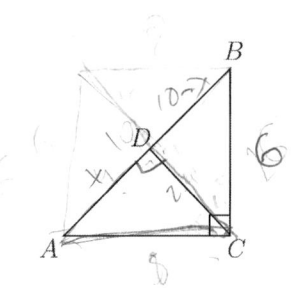

    A. 30 feet
    B. 40 feet
    C. 50 feet
    D. 60 feet
    E. 70 feet

50. In $\triangle ABC$ shown above, $\overline{CD}$ is drawn to $\overline{AB}$ so that $\overline{CD} \perp \overline{AB}$. If $AB = 10$ and $AC = 8$, what is the length of $\overline{CD}$?

    F. 2
    G. 2.4
    H. 4
    J. 4.8
    K. 6

STOP

SOLOMON ACADEMY — TEST 2 SOLUTIONS

Answers and Solutions
Practice Test 2

Answers

1. B	2. J	3. C	4. K	5. D
6. H	7. A	8. K	9. E	10. J
11. D	12. G	13. B	14. F	15. E
16. G	17. D	18. K	19. B	20. H
21. D	22. F	23. A	24. F	25. D
26. K	27. D	28. H	29. E	30. J
31. A	32. G	33. E	34. G	35. E
36. K	37. B	38. H	39. C	40. K
41. C	42. F	43. C	44. J	45. D
46. G	47. E	48. H	49. D	50. J

Solutions

1. (B)

$$5\frac{1}{2} \div 2\frac{3}{4} = \frac{11}{2} \div \frac{11}{4} = \frac{11}{2} \times \frac{4}{11} = 2$$

2. (J)

Tips
1. In scientific notation, all the numbers can be written in the form of $c \times 10^n$, where $1 \leq c < 10$ and n is an integer.
2. Use the property of exponent: $a^x \cdot a^y = a^{x+y}$

$$5 \times 10^3 \times 4.4 \times 10^2 = 5 \times 4.4 \times 10^3 \times 10^2$$
$$= 22 \times 10^5$$
$$= 2.2 \times 10 \times 10^5$$
$$= 2.2 \times 10^6$$

Since $10^3 \cdot 10^2 = 10^{2+3}$
$22 = 2.2 \times 10$

3. (C)

Tips Average $= \dfrac{\text{Sum of elements in a set}}{\text{number of elements}}$

Since the sum of the prices of gasoline is $3.25 + 3.75 + 3.5 + 3.9 = 14.4$, the average price of the gasoline is $\$\frac{14.4}{4}$ or $\$3.60$.

SOLOMON ACADEMY — Distribution or replication of any part of this page is prohibited. — TEST 2 SOLUTIONS

4. (K)

$3^3 = 27$, $3^2 = 9$, and $2^2 = 4$. Therefore, the value of $\frac{3^3 + 3^2}{2^2 + 2} = \frac{36}{6}$ or 6.

5. (D)

$$2x + 4 = y \qquad \text{Subtract 4 from each side}$$
$$2x = y - 4 \qquad \text{Divide each side by 2}$$
$$x = \frac{1}{2}y - 2$$

Therefore, x in terms of y is $\frac{1}{2}y - 2$.

6. (H)

Tips: Median is the middle number when a set of numbers in arranged from least to greatest. If there is n (odd number) of numbers in a set, the median is the middle number which is the $(\frac{n+1}{2})^{\text{th}}$ number in the set.

There are five numbers in set A. Thus, the median is the middle number, which is the third number in set A. Since $x < y < z < s < t$, z is the median of set A.

7. (A)

$$3(x - 2) + 5 = 23 \qquad \text{Subtract 5 from each side}$$
$$3(x - 2) = 18 \qquad \text{Divide each side by 3}$$
$$x - 2 = 6 \qquad \text{Add 2 to each side}$$
$$x = 8$$

Therefore, the value of x for which $3(x - 2) + 5 = 23$ is 8.

8. (K)

Substitute $2t$ for y in $2x - 4y - 12t = 0$ and solve for x.

$$2x - 4y - 12t = 0 \qquad \text{Substitute } 2t \text{ for } y$$
$$2x - 4(2t) - 12t = 0 \qquad \text{Simplify}$$
$$2x = 20t \qquad \text{Divide each side by 2}$$
$$x = 10t$$

Therefore, x in terms of t is $10t$.

9. (E)

Substitute -2 for x in $y = (x - 1)(x - 2)$ to evaluate the value of y.

$$y = (x - 1)(x - 2) \qquad \text{Substitute } -2 \text{ for } x$$
$$= (-2 - 1)(-2 - 2) \qquad \text{Simplify}$$
$$= 12$$

Therefore, the value of y is 12.

SOLOMON ACADEMY — TEST 2 SOLUTIONS

10. (J)

$\frac{2}{10} = 0.2$ and $\frac{3}{1000} = 0.003$. Therefore, the decimal representation of $10 + \frac{2}{10} + \frac{3}{1000} = 10.203$.

11. (D)

Since $2x = 3y = 6z = 12$, solve for x, y, and z for which $2x = 12$, $3y = 12$, and $6z = 12$, respectively. Thus, $x = 6$, $y = 4$ and $z = 2$. Therefore, the mean of x, y, and z is $\frac{x+y+z}{3} = \frac{6+4+2}{3}$ or 4.

12. (G)

There are four prime numbers less than 10: 2, 3, 5, and 7. Therefore, the sum of all prime numbers less than 10 is $2 + 3 + 5 + 7$ or 17.

13. (B)

Since 111 can be written as $111 = 13 \times 8 + 7$, the remainder is 7 when 111 is divided by 13.

14. (F)

$$\frac{12}{x} + 5 = 9 \quad \text{Subtract 5 from each side}$$
$$\frac{12}{x} = 4 \quad \text{Take the reciprocal of each side}$$
$$\frac{x}{12} = \frac{1}{4} \quad \text{Multiply each side by 12}$$
$$x = 3$$

Therefore, the value of x for which $\frac{12}{x} + 5 = 9$ is 3.

15. (E)

Tips: Use the property of exponents: $a^{x+y} = a^x \cdot a^y$

$$2^{x+3} = 2^x \times 2^3 \quad \text{Since } 2^x = b \text{ and } 2^3 = 8$$
$$= 8b$$

16. (G)

The pattern consists of F, G, H, J, and K. It is repeated continuously. Since every 5th is k, the 20th letter and 25th letter is also K. Therefore, the 26th letter is F and 27th letter is G.

17. (D)

Since $\sqrt{16} = 4$ and $\sqrt{49} = 7$,

$$(\sqrt{x} - \sqrt{y})^2 = (\sqrt{16} - \sqrt{49})^2 \quad \text{Substitute 16 for } x \text{ and 49 for } y$$
$$= (4 - 7)^2$$
$$= (-3)^2$$
$$= 9$$

SOLOMON ACADEMY — Distribution or replication of any part of this page is prohibited. — TEST 2 SOLUTIONS

18. (K)

Joshua has three suitcases that weigh 32 pounds each and two backpacks that weigh 16 pounds each. Therefore, the total weight of his luggage in pounds is $3(32) + 2(16) = 128$.

19. (B)

Tips   Distance formula: distance = rate × time

Use the distance formula. The total distance that car A travels in four hours is $65 \times 4 = 260$ miles.

20. (H)

Each face of the cube is a square. Since the area of each face is 16, the length of the cube is 4. Therefore, the volume of the cube is $4 \times 4 \times 4 = 64$.

21. (D)

There are 12 inches in one foot. Define x as the number of weeks that the tulip needs to grow 2 feet or $2 \times 12 = 24$ inches. Set up a proportion in terms of inches and weeks.

$$2_{\text{inches}} : 1_{\text{week}} = 24_{\text{inches}} : x_{\text{weeks}}$$
$$\frac{2}{1} = \frac{24}{x} \quad \text{Use cross product property}$$
$$2x = 24$$
$$x = 12$$

Therefore, it will take 12 weeks for the tulip to grow two feet.

22. (F)

$\overline{OC}$ and $\overline{OD}$ are the lines with a positive slope. Since $\overline{OC}$ is steeper (closer to y-axis) than $\overline{OD}$, $\overline{OC}$ has the larger value of slope. Therefore, $\overline{OC}$ has the largest value of slope.

23. (A)

Prime numbers are numbers that have only two factors: 1 and itself. There are 8 prime numbers less than twenty: 2, 3, 5, 7, 11, 13, 17, and 19.

24. (F)

$$x = y + 7 \quad \text{Switch sides}$$
$$y + 7 = x \quad \text{Subtract } x \text{ from each side}$$
$$y - x + 7 = 0 \quad \text{Subtract 7 to each side}$$
$$y - x = -7$$

Therefore, the value of $y - x$ is -7.

SOLOMON ACADEMY — TEST 2 SOLUTIONS

25. (D)

In the figure below, the trapezoid consists of two congruent 45°-45°-90° special right triangles, $\triangle ABC$ and $\triangle FED$, whose sides are in the ratio $1 : 1 : \sqrt{2}$. Since $AB = FE = 5\sqrt{2}$, $AC = BC = DE = DF = 5$.

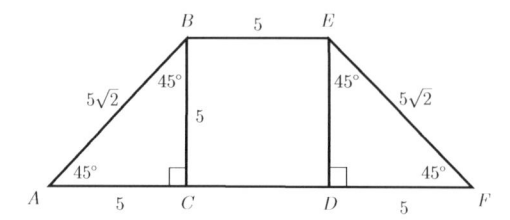

Triangles ABC and FED can be combined to create a square whose area is equal to the area of the square in the middle. The area of each square is 25. Therefore, the area of the trapezoid is 50.

26. (K)

$$AB + BC + CD = AD \quad \text{Substitute 9 for } AD \text{ and 4 for } BC$$
$$AB + CD + 4 = 9 \quad \text{Subtract 4 from each side}$$
$$AB + CD = 5$$

27. (D)

The perimeter of an equilateral triangle is the same as the perimeter of a square. Since the perimeter of the square with side length of 9 is 36, the perimeter of the equilateral triangle is also 36. Therefore, the length of the side of the equilateral triangle is $\frac{36}{3} = 12$.

28. (H)

> **Tips**: Given the two points (x_1, y_1) and (x_2, y_2), the midpoint formula is follows:
> $$\text{Midpoint} = \left(\frac{x_1 + x_2}{2}, \frac{y_1 + y_2}{2} \right)$$

To find the x and y coordinates of the midpoint between point $A(-1, 5)$ and point $B(3, -3)$, use the midpoint formula.

$$\text{Midpoint} = \left(\frac{x_1 + x_2}{2}, \frac{y_1 + y_2}{2} \right)$$
$$= \left(\frac{-1 + 3}{2}, \frac{5 + (-3)}{2} \right)$$
$$= (1, 1)$$

Therefore, the x and y coordinates of the midpoint between point A and point B is $(1, 1)$.

SOLOMON ACADEMY — TEST 2 SOLUTIONS

29. (E)

The clock indicates the first correct time at 12:12pm. Afterwards, it indicates correctly at 12:24pm, 12:36pm, 12:48pm and 1:00pm. This means that the clock only indicates 5 correct times every hour. Since there are five hours between 12:10pm to 5:10pm, the clock indicates time correctly $5 \times 5 = 25$ times. There are two additional times that the clock indicates correctly after 5pm: 5:12pm and 5:24pm. Therefore, the clock indicates time correctly 27 times from 12:10pm to 5:35pm.

30. (J)

At store A, two pens cost \$1.73. However, at store B, two pen cost \$1.99. Thus, the difference in the cost of two pens is \$1.99 − \$1.73 = \$0.26. Therefore, the amount you save per pen is $\frac{26}{2} = 13$ cents.

31. (A)

In the figure below, point X lies inside the square with side length of 9.

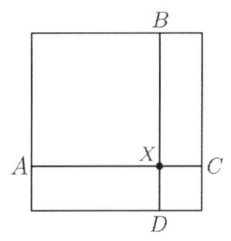

Since $\overline{AC}$ and $\overline{BD}$ are parallel to the sides of the square, both $XA + XC$ and $XB + XD$ equal the side length of the square, 9. Thus, $XA + XC = 9$ and $XB + XD = 9$. Therefore, $XA + XB + XC + XD = 18$.

32. (G)

$(x+2)^3 = 27$. Since $3^3 = 27$, $x + 2 = 3$. Thus, $x = 1$ and $x + 3 = 4$. Therefore, the value of $\frac{1}{x+3} = \frac{1}{4}$.

33. (E)

Answer choice (A) and (E) satisfy the Pythagorean theorem: $c^2 = a^2 + b^2$. However, $\sqrt{2}$ in answer choice (A) is not a positive integer. Thus, eliminate answer choice (A). Therefore, (E) is the correct answer.

34. (G)

$\overline{OB}$ is the longest line segment because it is the longest diagonal of the rectangular box. Therefore, (G) is the correct answer.

SOLOMON ACADEMY — TEST 2 SOLUTIONS

35. (E)

$AB = 6$. Thus, the area of the square is $6^2 = 36$. Since square $ABCD$ and triangle EFG have the same areas, the area of triangle EFG is 36. The base of triangle EFG is $\overline{EG}$ and $EG = 8$.

$$\text{Area of triangle } EFG = \frac{1}{2} \times (8) \times h$$
$$36 = 4 \times h$$
$$h = 9$$

Therefore, the height of triangle EFG is 9.

36. (K)

According to the survey, the most popular sport is the 100 meter sprint and the least popular sport is swimming.

$$\text{100 meter sprint} = 300 \times 0.36 = 108$$
$$\text{Swimming} = 300 \times 0.19 = 51$$

Therefore, $108 - 51 = 57$ million more people are going to watch the 100 meter sprint compared to swimming.

37. (B)

There are 12 inches in one foot. Convert Mr. Rhee's height to inches. Mr. Rhee is $5 \times 12 + 9 = 69$ inches tall. Jason is $\frac{1}{3} \times 69 = 23$ inches tall. Therefore, the difference in their heights in inches is $69 - 23 = 46$.

38. (H)

In order to make the equation true, the only possible value for x is zero when $y > 0$.

39. (C)

In the figure below, the area of each sector is the area of a quarter circle with a radius of 4 because its central angle is $90°$.

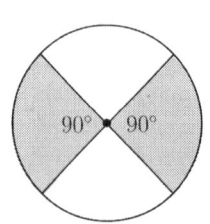

Thus, the sum of the area of the two shaded sectors equals the area of half of the circle. Therefore, the area of the shaded region is $\frac{1}{2}\pi(4)^2 = 8\pi$.

SOLOMON ACADEMY Distribution or replication of any part of this page is prohibited. TEST 2 SOLUTIONS

40. (K)

The units digit also means ones digit. The first five positive squares are 1, 4, 9, 16, and 25. Thus, the units digit of the first five positive squares are 1, 4, 9, 6, and 5, respectively.

$$\text{Sum of units digits} = 1 + 4 + 9 + 6 + 5 = 25$$

Therefore, the sum of the units digits of the first five positive squares is 25.

41. (C)

The base and height of $\triangle ABC$ are $2x - 2$ and x, respectively.

$$\text{Area of } \triangle ABC = \frac{1}{2}bh = \frac{1}{2}(2x-2)x$$
$$= x(x-1) = x^2 - x$$

Therefore, the area of $\triangle ABC$ is $A = x^2 - x$.

42. (F)

Distinct means different. There are five different positive integers in the set. It is necessary to solve this problem in terms of the sum: the sum is the average of the elements in the set times the number of the elements in the set. The average of the two smallest integers in the set is 2. This means that the sum of the two smallest integers is 4. Since the five integers in the set are positive and different, the two smallest positive integers in the set must be 1 and 3, neither 2 and 2, nor 0 and 4. Additionally, the averages of the three smallest integers, four smallest integers, and five integers in the set are 3, 4, and 5, respectively. Thus, the sums of the three smallest integers, four smallest integers and five integers in the set are 9, 16, and 25, respectively. Below shows how to obtain the five positive integers in the set.

$$\text{Sum of two smallest integers} = 1 + 3 = 4$$
$$\text{Sum of three smallest integers} = 1 + 3 + 5 = 9$$
$$\text{Sum of four smallest integers} = 1 + 3 + 5 + 7 = 16$$
$$\text{Sum of five integers in the set} = 1 + 3 + 5 + 7 + 9 = 25$$

Thus, there are 1, 3, 5, 7, and 9 in the set. Therefore, the largest integer in the set is 9.

43. (C)

In the figure below, two quadrilaterals $ABCD$ and $DEFG$ are parallelograms. A straight line passes through the points B, D, and F. This means that $\overline{BC} \parallel \overline{DE}$.

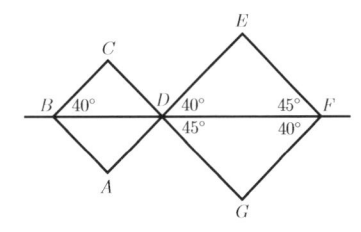

$\angle CBD \cong \angle EDF$ because they are corresponding angles. Thus, $m\angle CBD = m\angle EDF = 40°$. Use one of the properties of a parallelogram: opposite angles are congruent. In parallelogram $DEFG$, $\angle EDG$ and $\angle EFG$ are opposite angles. Since $m\angle EDG = 40° + 45° = 85°$, the measure of $\angle EFG$ is 85°.

SOLOMON ACADEMY — TEST 2 SOLUTIONS

44. (J)

In the figure below, Joshua is only allowed to walk either right or up, neither left nor down.

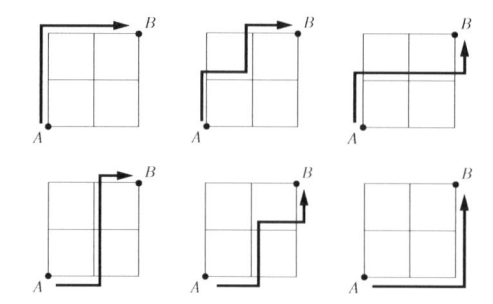

Therefore, there are six distinct 4-unit paths from point A to point B.

45. (D)

Tips: Use the difference of two squares pattern: $x^2 - y^2 = (x+y)(x-y)$.

$$x^2 - y^2 = 8 \qquad \text{Factor}$$
$$(x+y)(x-y) = 8 \qquad \text{Substitute 2 for } x-y$$
$$2(x+y) = 8$$
$$x + y = 4$$

Set up a system of equations: $x+y=4$ and $x-y=2$. Use the linear combinations method to solve for x and y.

$$\begin{aligned} x + y &= 4 \\ x - y &= 2 \qquad \text{Add two equations} \\ \hline 2x &= 6 \\ x &= 3 \end{aligned}$$

Substitute 3 for x in $x+y=4$ to solve for y.

$$x + y = 4 \qquad \text{Substitute 3 for } x$$
$$y + 3 = 4 \qquad \text{Subtract 3 from each side}$$
$$y = 1$$

Therefore, $x^2 + y^2 = (3)^2 + (1)^2 = 10$.

46. (G)

I is true: Both 9^2 and $(-9)^2$ equal 81.

II is false: The value of $\sqrt{4}$ is 2, not -2.

III is true: Any number divided by 0 is undefined.

IV is false: If $x < 0$, $|x| = -x$.

Thus, statements I and III are correct. Therefore, (B) is the correct answer.

SOLOMON ACADEMY   Distribution or replication of any part of this page is prohibited.   TEST 2 SOLUTIONS

47. (E)

> **Tips**   The fundamental counting principle: If one event can occur in m ways and another event can occur n ways, then the number of ways both events can occur is $m \times n$.

Mr. Rhee, Sue, Joshua, and Jason are taking a family photo. To count the total possible seating arrangements for Mr. Rhee's family, use the fundamental counting principle. Mr. Rhee has 4 ways to choose his seat out of four seats. After his seat is taken, Sue has 3 ways to choose her seat out of the three remaining seats. Joshua has 2 ways and Jason has 1 way. Thus, there are $4 \times 3 \times 2 \times 1 = 24$ possible seating arrangements for Mr.Rhee's family. Define R, S, A, and B as Mr. Rhee, Sue, Joshua, and Jason, respectively. To count the number of seating arrangements in which Mr. Rhee and Sue sit next to each other, let's consider two cases:

Case 1: Mr. Rhee sits left side of Sue. There are six possible arrangements as shown below.

$$\begin{array}{cccc} R & S & A & B \\ R & S & B & A \end{array} \qquad \begin{array}{cccc} A & R & S & B \\ B & R & S & A \end{array} \qquad \begin{array}{cccc} A & B & R & S \\ B & A & R & S \end{array}$$

Case 2: Mr. Rhee sits right side of Sue. There are six possible arrangements.

$$\begin{array}{cccc} S & R & A & B \\ S & R & B & A \end{array} \qquad \begin{array}{cccc} A & S & R & B \\ B & S & R & A \end{array} \qquad \begin{array}{cccc} A & B & S & R \\ B & A & S & R \end{array}$$

Thus, there are 12 possible seating arrangements in which Mr. Rhee sits next to Sue. Therefore, the probability that Mr. Rhee and Sue sit next to each other is $\frac{12}{24} = \frac{1}{2}$.

48. (H)

On Tuesday, students drank $\frac{1}{4}$ of the water. Thus, $\frac{3}{4}$ of the water would be remaining on Tuesday. Students drank $\frac{1}{3}$ of the remaining water on Wednesday. Thus, $\frac{2}{3}$ of the remaining water would be remaining on Wednesday. This means that $\frac{2}{3} \times \frac{3}{4} = \frac{1}{2}$ of the water would be remaining on Wednesday. Students drank $\frac{1}{2}$ of the remaining water on Thursday. Thus, $\frac{1}{2}$ of the remaining water would be remaining. Therefore, $\frac{1}{2} \times \frac{1}{2} = \frac{1}{4}$ of the water would be remaining on Thursday.

49. (D)

In the figure below, let x be the length of the shorter side of the rectangle. The rectangle is divided into two congruent smaller squares. Thus, the length of each square is x.

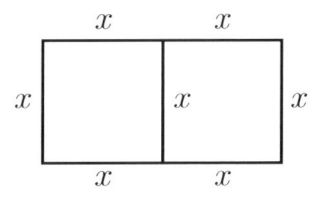

The total length of the fence can be expressed as $7x$ as shown above. Since the total length of the fence is 420 feet, $7x = 420 \Longrightarrow x = 60$. Therefore, the length of the shorter side of the rectangular field is 60.

50. (J)

> **Tips**: If a triangle is a right triangle, use the Pythagorean theorem: $c^2 = a^2 + b^2$, where c is the hypotenuse, a and b are the legs of the right triangle.

In the figure below, $\triangle ABC$ is a right triangle with $AB = 10$ and $AC = 8$. To find BC, use the Pythagorean theorem: $10^2 = 8^2 + BC^2$. Thus, $BC = 6$.

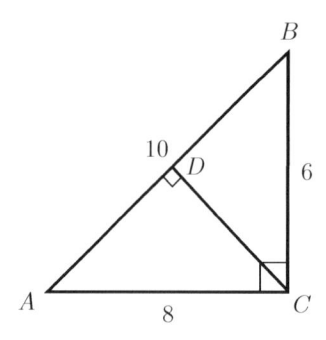

$\overline{AC} \perp \overline{BC}$. Consider $\overline{AC}$ and $\overline{BC}$ as the base and the height of $\triangle ABC$. Thus, the area of $\triangle ABC = \frac{1}{2} \times AC \times BC = \frac{1}{2}(8)(6) = 24$. Additionally, $\overline{CD} \perp \overline{AB}$. If you consider $\overline{AB}$ and $\overline{CD}$ as the base and the height of $\triangle ABC$, the area of $\triangle ABC = \frac{1}{2} \times AB \times CD$.

$$\text{Area of } \triangle ABC = \frac{1}{2} \times AB \times CD = 24 \quad \text{Substitute 10 for } AB$$
$$\frac{1}{2}(10)\, CD = 24 \quad \text{Solve for } CD$$
$$CD = 4.8$$

SOLOMON ACADEMY

PRACTICE TEST 3
MATHEMATICS PROBLEMS
50 Questions
Time — 60 minutes

Directions: Solve each problem and enter your answer by marking the circle on the answer sheet. Choose the best answer among the answer choices given.

1. $1\frac{4}{5} \div 0.6 =$

    A. 1
    B. 2
    C. 3
    D. 4
    E. 5

2. How many prime factors does 30 have?

    F. 3
    G. 4
    H. 5
    J. 6
    K. 7

3. Evaluate $x^2 - 3x - 9$ when $x = -3$.

    A. 9
    B. 1
    C. 0
    D. -9
    E. -27

4. Jason scored 94 and 88 on two math tests. If he wants to get an average of 92 for all three math tests, what score does he need to get on the third math test?

    F. 91
    G. 92
    H. 93
    J. 94
    K. 95

5. Solve for x: $\frac{4}{5}x + 2 = 10$.

    A. 12
    B. 10
    C. 8
    D. 6
    E. 5

6. Which of the following expression has the largest value?

    F. $\sqrt{100}$
    G. $(-2)^3$
    H. $|-12|$
    J. $\frac{1}{\frac{1}{13}}$
    K. $(-3)^2$

7. A set has the following five distinct integers: 5, 6, 10, 13, and x. What is the largest possible integer value of x so that x is the median of the set?

   A. 6
   B. 7
   C. 8
   D. 9
   E. 10

8. If a car travels 13 miles in 15 minutes, how fast, in miles per hour, does the car travel?

   F. 26
   G. 39
   H. 52
   J. 60
   K. 65

9. Which of the following number has only four factors?

   A. 15
   B. 18
   C. 20
   D. 24
   E. 30

10. If $3x = 6y - 9$, what is y in terms of x?

    F. $3x - 9$
    G. $3x + 9$
    H. $\frac{3x-9}{6}$
    J. $\frac{1}{2}x + \frac{2}{3}$
    K. $\frac{1}{2}x + \frac{3}{2}$

11. Joshua has three more than twice the number of books that Jason has. If Joshua has 17 books, how many books does Jason have?

    A. 7
    B. 8
    C. 9
    D. 10
    E. 11

12. A pipe was cut into four smaller pipes: 24 inches, 20 inches, 16 inches, and 36 inches. How long, in feet, was the original pipe?

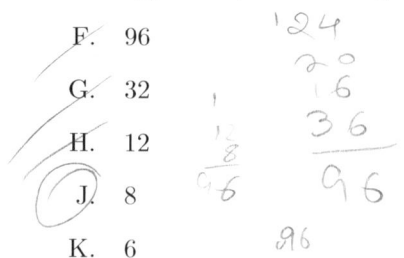

    F. 96
    G. 32
    H. 12
    J. 8
    K. 6

13. Let A be the area of a triangle. If the lengths of the base and height are doubled and tripled, respectively, what is the area of a new triangle in terms of A?

    A. A
    B. $2A$
    C. $3A$
    D. $6A$
    E. $12A$

14. $\frac{2}{x} + \frac{3}{x} = 1$, what is the value of x?

    F. 1
    G. 5
    H. 10
    J. 15
    K. 20

15. $2x + 3y = 24$. If $x = 3t$ and $y = 2t$, what is the value of t?

    A. 8
    B. 6
    C. 5
    D. 4
    E. 2

16. If k is divided by 4, the quotient is 7 and remainder is 3. What is the value of k?

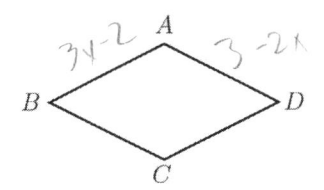

    F. 84
    G. 52
    H. 31
    J. 25
    K. 19

17. The quadrilateral shown above is a rhombus. If $AB = 3x - 2$ and $AD = 3 - 2x$, what is the value of x?

    A. 1
    B. 2
    C. 3
    D. 4
    E. 5

18. Mr. Rhee bought 3 oranges at $1.49 each and 4 apples at $1.29 each at the store yesterday. The store goes on sales today. The new price of each orange and apple are $1.19 and $1.09, respectively. How much would Mr. Rhee have saved if he bought the same number of fruits at the store today?

    F. $0.50
    G. $0.90
    H. $1.40
    J. $1.50
    K. $1.70

19. Which of the following expression is equal to the sum of x and $2x$ is multiplied by $3x$?

    A. $9x^3$
    B. $5x^3$
    C. $5x^2$
    D. $6x^2$
    E. $9x^2$

20. Which of the following fraction is the largest in value?

    F. $\frac{1}{3}$ divided by $\frac{1}{5}$
    G. $\frac{1}{3}$ divided by $\frac{1}{4}$
    H. $\frac{1}{4}$ divided by $\frac{1}{5}$
    J. $\frac{1}{4}$ divided by $\frac{1}{6}$
    K. $\frac{1}{5}$ divided by $\frac{1}{7}$

21. Three points A, B, and C lie on the same number line, not necessarily in that order. Point A and point B are 50 units apart and point C and point A are 17 units apart. What is the maximum distance that point C and point B are apart?

   A. 17
   B. 33
   C. 50
   D. 67
   E. 80

22. The post office charges \$2.55 for the first ounce and \$1.65 for each additional ounce of a package. If the weight of the package is 11 ounces, which of the following expression represents the total cost, in dollars, of the package?

   F. $2.55 + 1.65$
   G. $1.65 + 2.55(10)$
   H. $1.65 + 2.55(11)$
   J. $2.55 + 1.65(10)$
   K. $2.55 + 1.65(11)$

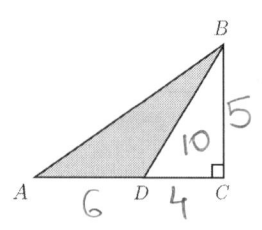

23. In the figure above, $AD = 6$, $DC = 4$, and the area of $\triangle ABC = 25$. What is the area of the shaded region?

   A. 14
   B. 15
   C. 16
   D. 17
   E. 18

24. Which of the following number has a remainder of 2 if the number is divided by the sum of its digits?

   F. 12
   G. 15
   H. 23
   J. 26
   K. 36

25. x is twice of y and y is two more than three times z. If $z = 2$, what is the value of x?

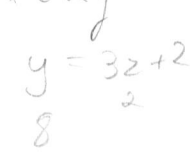

   A. 13
   B. 14
   C. 15
   D. 16
   E. 17

x	1	2	3	4	5
y	4	7	10	13	

26. If the pattern shown on the table above continues, what is the value of y when $x = 5$?

   F. 20
   G. 19
   H. 18
   J. 17
   K. 16

27. Jason and Joshua ordered a pizza and ate one-third of it. One hour later, they ate half of the remaining part of the pizza. What fractional part of the pizza have Jason and Joshua eaten?

    A. $\frac{1}{4}$
    B. $\frac{1}{3}$
    C. $\frac{1}{2}$
    D. $\frac{3}{5}$
    E. $\frac{2}{3}$

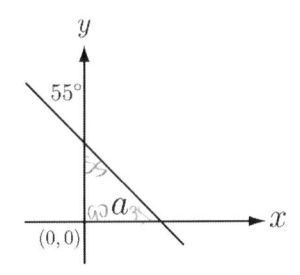

28. In the graph above, a line passes through the x and y axis. What is the measure of $\angle a$?

    F. 25
    G. 35
    H. 45
    J. 55
    K. 65

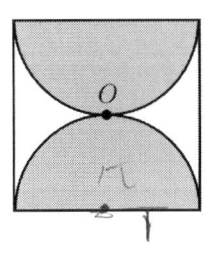

29. In the figure above, Point O is the center of the square with side length of 2. Two semicircles are tangent to each other inside the square. What is the area of the shaded region?

    A. $\frac{\pi}{2}$
    B. π
    C. 2π
    D. 3π
    E. 4π

30. The average of the three numbers, 2, 3, and $x - 1$ is n. What is the value of x in terms of n ?

    F. $n + 4$
    G. $n - 4$
    H. $3n - 2$
    J. $3n - 4$
    K. $3n - 6$

31. If 100 is divided by k, the quotient is 33 and the remainder is r. Which of the following expression is equal to 100?

    A. $33r + k$
    B. $33k + r$
    C. $33 + kr$
    D. $kr + 33$
    E. $k + r + 33$

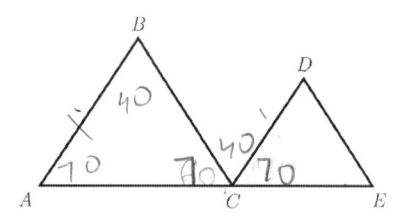

32. In the figure above, $\overline{AB} \parallel \overline{CD}$ and $\overline{BC} \parallel \overline{DE}$. The three points, A, C, and E lie on the same line. If $m\angle BAC = 70°$ and $m\angle ABC = 40°$, what is the measure of $\angle BCD$?

   F. 20
   G. 30
   H. 40
   J. 50
   K. 60

$$1 \leq 2x + 1 \leq 11$$

33. Set S consists of integers that satisfy the inequality above. How many elements does set S have?

   A. 2
   B. 3
   C. 4
   D. 5
   E. 6

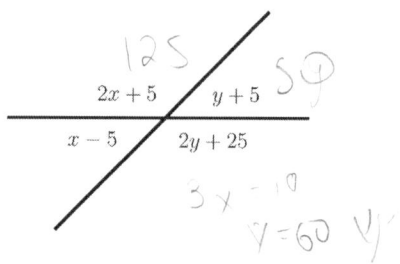

34. Two lines intersect and form two pairs of vertical angles. What is the value of $x + y$?

   F. 80
   G. 90
   H. 100
   J. 110
   K. 120

35. How many positive integers less than 100 are divisible by 2 and 5?

   A. 10
   B. 9
   C. 8
   D. 7
   E. 6

36. The sum of x and y is 9. If $y = 4x - 1$, what is the value of x?

   F. 2
   G. 3
   H. 4
   J. 5
   K. 6

37. Jason deposited $129 into his savings account. He withdrew $27 from his savings account to pay for a calculator. He wants to buy as many books as possible with the remaining money in his account. What is the maximum number of books he can buy if each book costs $11?

   A. 7
   B. 8
   C. 9
   D. 10
   E. 11

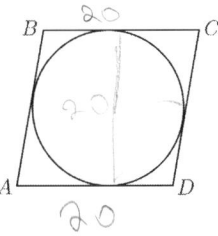

38. In the figure above, a circle is inscribed in parallelogram $ABCD$. The diameter of the circle is the height of the parallelogram. If $AD = 20$ and the radius of the circle is 10, what is the area of parallelogram $ABCD$?

   F. 200
   G. 300
   H. 400
   J. 500
   K. 600

39. There are three books on a bookshelf: two math books and a history book. If you arrange these three books, how many different arrangements are possible?

   A. 27
   B. 9
   C. 6
   D. 3
   E. 1

40. What is the arithmetic mean of the first five nonnegative even integers?

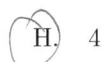

   F. 2
   G. 3
   H. 4
   J. 5
   K. 6

Bicycling	♡ ♡
Running	♡ ♡ ♡ ♡
Rowing	♡ ♡
Swimming	♡

♡ = 1750 Calories

41. Mr. Rhee is going to a gym to lose weight. The table above summarizes the types of cardiovascular exercises he had done in May. Assuming 3500 calories is equal to 1 pound of body fat, how many pounds of body fat did Mr. Rhee lose in May?

   A. 5
   B. 6
   C. 7
   D. 8
   E. 9

42. Which of the following expression is equal to $\frac{b+a}{ab}$?

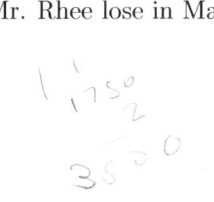

   F. 2
   G. $a+b$
   H. $\frac{1}{a} + \frac{1}{b}$
   J. $\frac{1}{a+b}$
   K. $\frac{a+1}{a}$

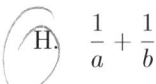

43. In a circular track whose circumference is 60 m, Joshua and Jason are running in opposite directions from a starting position. Joshua is running 2 m/s and Jason is running 1 m/s. By the time Joshua and Jason meet each other for the first time, how many meters did Joshua run?

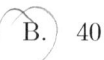

   A. 45
   B. 40
   C. 36
   D. 30
   E. 24

44. All of the numbers have the same number of factors except which of the following?

   F. 4
   G. 9
   H. 25
   J. 49
   K. 51

45. If a car factory produces x cars in y months, how many cars does the car factory produce in z years?

   A. $\dfrac{xz}{y}$
   B. $\dfrac{xy}{z}$
   C. $\dfrac{12yz}{x}$
   D. $\dfrac{12xz}{y}$
   E. $\dfrac{12xy}{z}$

46. On the number line above, a total of nine tick marks will be placed from $\sqrt{0.81}$ to $\sqrt{6.25}$ inclusive. If all the tick marks are evenly spaced, what is the length between any consecutive tick marks?

   F. 0.16
   G. 0.18
   H. 0.20
   J. 0.22
   K. 0.24

47. In the table above, positive integers from 1 to 9 are arranged such that the sum of the numbers in any horizontal, vertical, or main diagonal line is always the same number. Which of the following is the value of x in the table?

   A. 9
   B. 7
   C. 3
   D. 2
   E. 1

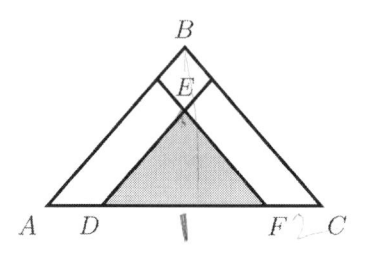

48. In the figure above, $\overline{AB} \parallel \overline{DE}$ and $\overline{BC} \parallel \overline{EF}$. If the ratio of DF to AC is 1 to 2, what is the ratio of the area of the shaded region to that of the unshaded region?

F. $\dfrac{1}{3}$

G. $\dfrac{2}{5}$

H. $\dfrac{1}{2}$

J. $\dfrac{3}{5}$

K. $\dfrac{2}{3}$

49. Two standard dice are rolled. What is the probability that the product of the two numbers shown on top of the two dice is odd?

A. $\dfrac{3}{4}$

B. $\dfrac{2}{3}$

C. $\dfrac{1}{2}$

D. $\dfrac{1}{3}$

E. $\dfrac{1}{4}$

50. An indoor swimming pool opens at 9 am and children start swimming. Staff at the swimming pool strictly follow a safety rule such that children must take a 10 minute break every 45 minutes. According to the safety rule, children take the first break time starting at 9:45 am until 9:55 am. If the swimming pool closes at 6 pm, at what time would the last break time begin?

F. 5:00 pm

G. 5:05 pm

H. 5:10 pm

J. 5:15 pm

K. 5:20 pm

STOP

SOLOMON ACADEMY — TEST 3 SOLUTIONS

Answers and Solutions
Practice Test 3

Answers

1. C	2. F	3. A	4. J	5. B
6. J	7. D	8. H	9. A	10. K
11. A	12. J	13. D	14. G	15. E
16. H	17. A	18. K	19. E	20. F
21. D	22. J	23. B	24. J	25. D
26. K	27. E	28. G	29. B	30. J
31. B	32. H	33. E	34. J	35. B
36. F	37. C	38. H	39. D	40. H
41. A	42. H	43. B	44. K	45. D
46. H	47. A	48. F	49. E	50. G

Solutions

1. (C)

   Since $1\frac{4}{5} = \frac{9}{5}$ and $0.6 = \frac{3}{5}$,

   $$1\frac{4}{5} \div 0.6 = \frac{9}{5} \div \frac{3}{5} = \frac{9}{5} \times \frac{5}{3} = 3$$

2. (F)

   The factors of 30 are 1, 2, 3, 5, 6, 10, 15, and 30. Therefore, there are three prime factors of 30: 2, 3, and 5.

3. (A)

   $$\begin{aligned} x^2 - 3x - 9 &= (-3)^2 - 3(-3) - 9 \quad &\text{Substitute } -3 \text{ for } x \\ &= 9 + 9 - 9 \\ &= 9 \end{aligned}$$

   Therefore, the value of $x^2 - 3x - 9$ when $x = -3$ is 9.

4. (J)

   > Tips: Sum of elements in a set = average of elements × number of elements

   Since the average of the three tests is 92, the sum of the three tests is $92 \times 3 = 276$. Therefore, the score that Jason needs to get on the third test is $276 - 94 - 88$ or 94.

SOLOMON ACADEMY — TEST 3 SOLUTIONS

5. (B)

$$\frac{4}{5}x + 2 = 10 \quad \text{Subtract 2 from each side}$$
$$\frac{4}{5}x = 8 \quad \text{Multiply each side by } \frac{5}{4}$$
$$\frac{5}{4} \cdot \frac{4}{5}x = \frac{5}{4}(8)$$
$$x = 10$$

Therefore, the value of x for which $\frac{4}{5}x + 2 = 10$ is 10.

6. (J)

F. $\sqrt{100} = 10$

G. $(-2)^3 = -8$

H. $|-12| = 12$

J. $\dfrac{1}{\frac{1}{13}} = 13$

K. $(-3)^2 = 9$

Therefore, the expression in answer choice (J) has the largest value.

7. (D)

Since x is the median of the set, rearrange the five distinct integers from least to greatest: 5, 6, x, 10, and 13. In order for x to be the median, the possible integer values for x are 7, 8, and 9. Therefore, the largest possible integer value of x is 9.

8. (H)

A car travels 13 miles in 15 minutes. This means that the car travels 13 miles every 15 minutes. Since there are 60 minutes in one hour, the car travels 13×4 or 52 miles in one hour. Therefore, the car travels at the rate of 52 miles per hour.

9. (A)

15 has 4 factors: 1, 3, 5, and 15. Both 18 and 20 have 6 factors. Both 24 and 30 have 8 factors. Therefore, (A) is the correct answer.

10. (K)

$$6y - 9 = 3x \quad \text{Add 9 to each side}$$
$$6y = 3x + 9 \quad \text{Divide each side by 6}$$
$$y = \frac{3}{6}x + \frac{9}{6} \quad \text{Simplify}$$
$$y = \frac{1}{2}x + \frac{3}{2}$$

SOLOMON ACADEMY — TEST 3 SOLUTIONS

11. (A)

Let x be the number of books that Jason has. Then, the number of books that Joshua has can be written as $2x + 3$. Since Joshua has 17 books, $2x + 3 = 17$.

$$2x + 3 = 17 \quad \text{Subtract 3 from each side}$$
$$2x = 14 \quad \text{Divide each side by 2}$$
$$x = 7$$

Therefore, Jason has 7 books.

12. (J)

The sum of the lengths of the smaller pipes is $24 + 20 + 16 + 36$ or 96 inches. Since there are 12 inches in one foot, 96 inches is equivalent to 8 feet. Therefore, the length of the original pipe is 8 feet.

13. (D)

Let b and h be the lengths of the base and height of the triangle. Since the length of the base is doubled and the length of the height is tripled, the lengths of the new base and new height can be written as $2b$ and $3h$, respectively. Thus,

$$\text{Area of a new triangle} = \frac{1}{2}(2b)(3h)$$
$$= 6 \times \frac{1}{2}bh \quad \text{Since } A = \frac{1}{2}bh$$
$$= 6A$$

Therefore, the area of the new triangle in terms of A is $6A$.

14. (G)

> Tips   Use the distributive property: $a(b + c) = ab + ac$

$$\frac{2}{x} + \frac{3}{x} = 1 \quad \text{Multiply each side by } x$$
$$x\left(\frac{2}{x} + \frac{3}{x}\right) = x(1) \quad \text{Use the distributive property}$$
$$2 + 3 = x$$
$$x = 5$$

Therefore, the value of x for which $\frac{2}{x} + \frac{3}{x} = 1$ is 5.

15. (E)

Since $x = 3t$ and $y = 2t$, substitute $3t$ for x and $2t$ for y in $2x + 3y = 24$.

$$2x + 3y = 24 \qquad \text{Substitute } 3t \text{ for } x \text{ and } 2t \text{ for } y$$
$$2(3t) + 3(2t) = 24$$
$$12t = 24$$
$$t = 2$$

Therefore, the value of t is 2.

16. (H)

> Tips: When k is divided by p, the quotient is q and the remainder is r. Then, k can be written as $k = pq + r$

When k is divided by 4, the quotient is 7 and the remainder is 3. Therefore, $k = 4 \times 7 + 3$, or 31.

17. (A)

A rhombus is a quadrilateral with four equal sides. Thus, set AB equal to AD and solve for x.

$$AB = AD$$
$$3x - 2 = 3 - 2x$$
$$5x = 5$$
$$x = 1$$

Therefore, the value of x is 1.

18. (K)

Mr. Rhee would have saved $0.30 per orange and $0.20 per apple if he went to the store today.

$$\text{Savings} = 3(1.49 - 1.19) + 4(1.29 - 1.09)$$
$$= 3(0.3) + 4(0.2)$$
$$= \$1.70$$

Therefore, Mr. Rhee would have saved $1.70 if he went to the store today.

19. (E)

Translate the verbal phrase into a mathematical expression. The sum of x and $2x$ can be expressed as $x + 2x$. Thus,

$$3x(x + 2x) = 3x(3x)$$
$$= 9x^2$$

Therefore, the sum of x and $2x$ is multiplied by $3x$ is equal to $9x^2$.

20. (F)

F. $\dfrac{1}{3} \div \dfrac{1}{5} = \dfrac{1}{3} \times \dfrac{5}{1} = \dfrac{5}{3}$ \qquad G. $\dfrac{1}{3} \div \dfrac{1}{4} = \dfrac{1}{3} \times \dfrac{4}{1} = \dfrac{4}{3}$

H. $\dfrac{1}{4} \div \dfrac{1}{5} = \dfrac{1}{4} \times \dfrac{5}{1} = \dfrac{5}{4}$ \qquad J. $\dfrac{1}{4} \div \dfrac{1}{6} = \dfrac{1}{4} \times \dfrac{6}{1} = \dfrac{3}{2}$

K. $\dfrac{1}{5} \div \dfrac{1}{7} = \dfrac{1}{5} \times \dfrac{7}{1} = \dfrac{7}{5}$

Therefore, (F) is the correct answer.

21. (D)

Place point A in the middle on the number line as shown in figure 1. Since points A, B and C lie on the same number line, not necessarily in that order, place point B 50 units left of point A or 50 units right of point A.

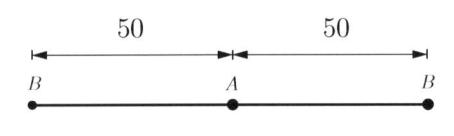

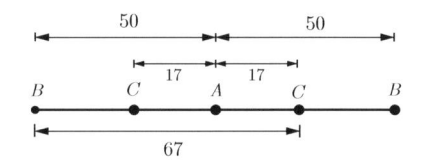

Figure 1 \qquad Figure 2

Point C and point A are 17 units apart. So, place point C 17 units left of point A or 17 units right of point A as shown in figure 2. Therefore, the maximum distance that point C and point B are apart is 67 units.

22. (J)

The first ounce costs $2.55 and each additional ounce costs $1.65. If a package weighs 11 ounces, the first ounce will cost $2.55 and the additional 10 ounces will cost $1.65 × 10. Therefore, the total cost of the package, in dollars, is $2.55 + 1.65(10)$.

23. (B)

Use the area $A = \frac{1}{2}bh$ formula to find the height of $\triangle ABC$. Since the area of $\triangle ABC$ is 25 and $AC = AD + DC = 10$, the height, BC, is 5.

$$\text{Area of the shaded region} = \text{Area of } \triangle ABD$$
$$= \frac{1}{2}(AD)(BC)$$
$$= \frac{1}{2}(6)(5)$$
$$= 15$$

Therefore, the area of the shaded region is 15.

SOLOMON ACADEMY — TEST 3 SOLUTIONS

24. (J)

F. $12 \div (1+2) \implies$ remainder of 0
G. $15 \div (1+5) \implies$ remainder of 3
H. $23 \div (2+3) \implies$ remainder of 3
J. $26 \div (2+6) \implies$ remainder of 2
K. $36 \div (3+6) \implies$ remainder of 0

25. (D)

Since $z = 2$, $y = 3z + 2 = 8$. Therefore, $x = 2y = 16$.

26. (K)

The table shows that as the value of x is increased by 1, the value of y is increased by 3. Therefore, the value of y when $x = 5$ is $13 + 3 = 16$.

27. (E)

Jason and Joshua ate one-third of the pizza. So, the remaining part of the pizza is two-thirds of the pizza. One hour later, they ate half of the remaining pizza. This means that they ate $\frac{1}{2} \times \frac{2}{3} = \frac{1}{3}$ of the pizza one hour later. Therefore, the fractional part of the pizza they have eaten is $\frac{1}{3} + \frac{1}{3} = \frac{2}{3}$ pizza.

28. (G)

In the figure below, $m\angle b = 55°$ because it is a vertical angle.

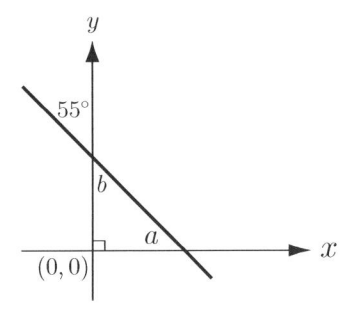

$\angle a$ and $\angle b$ are complementary angles whose sum of their measures is $90°$. Therefore, the measure of $\angle a$ is $35°$.

29. (B)

The shaded region consists of two semicircles tangent to each other. Thus, the area of the shaded region equals the area of one circle with a radius of 1. Therefore, the area of the shaded region is $\pi(1)^2 = \pi$.

SOLOMON ACADEMY — TEST 3 SOLUTIONS

30. (J)

> **Tips** Sum of elements in a set = average of elements × number of elements

Often, when solving problems regarding average, it is easier to solve the problems in terms of the sum.

$$\text{Sum} = \text{Average} \times 3$$
$$2 + 3 + x - 1 = 3n$$
$$x + 4 = 3n$$
$$x = 3n - 4$$

Therefore, the value of x in terms of n is $3n - 4$.

31. (B)

> **Tips** When k is divided by p, the quotient is q and the remainder is r. Then, k can be written as $k = pq + r$

If 100 is divided by k, the quotient is 33 and the remainder is r. Therefore, 100 can be written as $100 = 33k + r$.

32. (H)

In the figure below, $\overline{AB} \parallel \overline{CD}$. $\angle BAC$ and $\angle DCE$ are corresponding angles so that $m\angle BAC = m\angle DCE = 70°$.

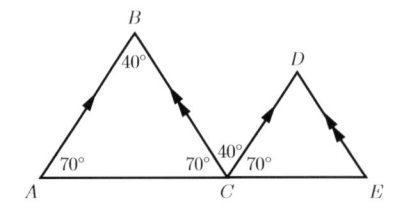

Since $m\angle ABC = 40°$, $m\angle BCA = 70°$ because the sum of the measures of interior angles of a triangle is 180°. The straight angle at C is 180°. Therefore, $m\angle BCD = 40°$.

33. (E)

$$1 \leq 2x + 1 \leq 11 \qquad \text{Subtract 1 from each side}$$
$$0 \leq 2x \leq 10 \qquad \text{Divide each side by 2}$$
$$0 \leq x \leq 5$$

Thus, the values of x for which $0 \leq x \leq 5$ are 0, 1, 2, 3, 4, and 5. Therefore, set S has six elements.

SOLOMON ACADEMY — TEST 3 SOLUTIONS

34. (J)

> **Tips** Supplementary angles are two angles whose sum of their measures is 180°.

$2x + 5$ and $x - 5$ are supplementary angles. Thus, the sum of their measures is 180°.

$$2x + 5 + x - 5 = 180$$
$$3x = 180$$
$$x = 60$$

Additionally, $y + 5$ and $2y + 25$ are supplementary angles.

$$y + 5 + 2y + 25 = 180$$
$$3y + 30 = 180$$
$$y = 50$$

Therefore, $x + y = 60 + 50 = 110$.

35. (B)

If a number is divisible by 2 and 5, the number must be divisible by 10. Since the positive integers from 1 to 99 are under consideration, there are 9 positive integers that are divisible by 10: $10, 20, \cdots, 80$, and 90.

36. (F)

The sum of x and y is 9. This can be expressed as $x + y = 9$. Since $y = 4x - 1$, substitute $4x - 1$ for y in $x + y = 9$ and solve for x.

$$x + y = 9 \quad\quad \text{Substitute } 4x - 1 \text{ for } y$$
$$x + 4x - 1 = 9$$
$$5x = 10$$
$$x = 2$$

Therefore, the value of x is 2.

37. (C)

Jason spent \$27 on a calculator. Thus, he has a remaining balance of \$129 − \$27 = \$102 in his savings account. Each book costs \$11. Jason cannot buy ten books because the cost for ten books is \$110, which exceeds \$102. Therefore, the maximum number of books that Jason can buy is 9.

38. (H)

> **Tips** Area of parallelogram is $A = bh$, where b is the length of the base and h is the length of the height.

Since the radius of the circle is 10, the diameter of the circle is 20 which is also the height of the parallelogram. $\overline{AD}$ is the base of the parallelogram and $AD = 20$. Therefore, the area of the parallelogram is $bh = 20 \times 20 = 400$.

SOLOMON ACADEMY   Distribution or replication of any part of this page is prohibited.   TEST 3 SOLUTIONS

39. (D)

The list below shows the possible arrangements for the two math books and the history book (M and H are used for the math book and the history book, respectively).

$$M \quad M \quad H$$
$$M \quad H \quad M$$
$$H \quad M \quad M$$

Therefore, there are three possible arrangements for the two math books and the history book.

40. (H)

The first five nonnegative even integers are 0, 2, 4, 6, and 8.

$$\text{Arithmetic mean} = \frac{0+2+4+6+8}{5} = \frac{20}{5} = 4$$

Therefore, the arithmetic mean of the first five nonnegative even integers is 4.

41. (A)

According to the table, the total amount of calories that Mr. Rhee lost in May is $10 \times 1750 = 17500$. Since 3500 calories is equal to 1 pound of body fat,

$$\text{Total pounds of body fat lost} = \frac{17500}{3500} = 5$$

Therefore, Mr. Rhee lost 5 pounds of body fat in May.

42. (H)

$$\frac{b+a}{ab} = \frac{b}{ab} + \frac{a}{ab}$$
$$= \frac{1}{a} + \frac{1}{b}$$

43. (B)

> Tips   Use the distance formula: distance = rate × time.

Joshua is running 2 m/s and Jason is running 1 m/s. Define t, in seconds, as how long Joshua and Jason have been running until they meet each other for the first time. Then, the distance that Joshua and Jason run for t seconds are $2t$ and t, respectively. When Joshua and Jason meet each other, the sum of the distance that both run equals to the circumference of the circular track, 60 m. Thus,

$$2t + t = 60$$
$$3t = 60$$
$$t = 20$$

This means that Joshua runs for 20 seconds. Therefore, Joshua runs $20 \times 2 = 40$ meters.

SOLOMON ACADEMY — TEST 3 SOLUTIONS

44. (K)

Determine the number of factors in each answer choice.

F. $4 = \{1, 2, 4\}$ G. $9 = \{1, 3, 9\}$
H. $25 = \{1, 5, 25\}$ J. $49 = \{1, 7, 49\}$
K. $51 = \{1, 3, 17, 51\}$

All of the answer choices have three factors except answer choice (K). Therefore, (K) is the correct answer.

45. (D)

There are 12 months in 1 year. Convert z years to $12z$ months. Define p as the number of cars that the car factory produce in $12z$ months. Set up a proportion in terms of cars and months.

$$x \text{ cars} : y \text{ months} = p \text{ cars} : 12z \text{ months}$$
$$\frac{x}{y} = \frac{p}{12z} \quad \text{Use cross product property}$$
$$py = 12xz$$
$$p = \frac{12xz}{y}$$

Therefore, the number of cars that the car factory produce in z years is $\frac{12xz}{y}$.

46. (H)

There are nine tick marks between $\sqrt{0.81} = 0.9$ and $\sqrt{6.25} = 2.5$ inclusive. Let's define interval as the length between any consecutive tick marks. There is one interval between two consecutive tick marks. There are two intervals between three consecutive tick marks, three intervals between four consecutive tick marks, and so on so forth. If the pattern continues, there are eight intervals between nine consecutive tick marks.

$$\text{Length of one interval} = \frac{\text{Distance between 0.9 and 2.5}}{8 \text{ intervals}} = \frac{1.6}{8} = 0.2$$

Therefore, the length between any consecutive tick marks is 0.2.

47. (A)

Find the sum of the numbers in any horizontal, vertical, or main diagonal. In table 1, there are three numbers on the diagonal so the sum is $4 + 5 + 6 = 15$.

	6	
x	5	
4		8

Table 1

	6	
x	5	1
4		8

Table 2

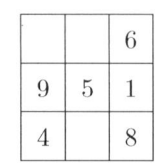

	6	
9	5	1
4		8

Table 3

Place 1 on the third column so that the sum is 15 as shown in table 2. Since the sum of the numbers in the second row is 15, $x + 5 + 1 = 15$. Therefore, the value of x is 9 as shown in table 3.

48. (F)

> **Tips:** If two triangles are similar, the ratio of their areas is equal to the square of the ratio of any pair of corresponding sides.

In the figure below, $\overline{AB} \parallel \overline{DE}$ and $\overline{BC} \parallel \overline{EF}$. This implies that $\angle A \cong \angle D$ and $\angle C \cong \angle F$ because they are corresponding angles. Since the two interior angles of $\triangle ABC$ and $\triangle DEF$ are congruent, $\angle B \cong \angle E$. Thus, $\triangle ABC$ and $\triangle DEF$ are similar triangles.

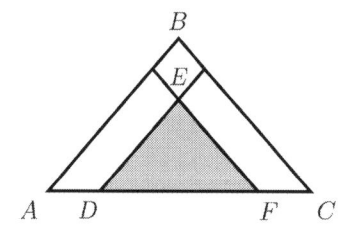

Since the ratio of DF to AC is 1 to 2, the ratio of their areas is equal to the square of the ratio of of any pair of corresponding sides.

$$\frac{\text{Area of } \triangle DEF}{\text{Area of } \triangle ABC} = \left(\frac{DF}{AC}\right)^2 = \left(\frac{1}{2}\right)^2 = \frac{1}{4}$$

Since the ratio of their areas is 1 to 4, define x as the area of $\triangle DEF$ and $4x$ as the area of $\triangle ABC$. Then, the area of the unshaded region equals the area of $\triangle ABC$ minus the area of $\triangle DEF$, which can be expressed as $4x - x$ or $3x$.

$$\frac{\text{Area of shaded region}}{\text{Area of unshaded region}} = \frac{x}{3x} = \frac{1}{3}$$

Therefore, the ratio of the area of the shaded region to that of the unshaded region is $\frac{1}{3}$.

49. (E)

> **Tips:** The fundamental counting principle: If one event can occur in m ways and another event can occur n ways, then the number of ways both events can occur is $m \times n$.

Use the fundamental counting principle. Define event 1 and event 2 as selecting a number from the first and second die, respectively. There are six outcomes from each event. Thus, the total number of outcomes for event 1 and event 2 is $6 \times 6 = 36$. Define event 3 and event 4 as selecting an odd number from the first and second die, respectively. There are three outcomes from each event: 1, 3, and 5. Thus, the total number of outcomes for event 3 and event 4 is $3 \times 3 = 9$. Thus,

$$\text{Probability that the product is odd} = \frac{9}{36} = \frac{1}{4}$$

Therefore, the probability that the product of the two numbers is odd is $\frac{1}{4}$.

50. (G)

Table 1 below shows the first couple of times swimming sessions and mandatory breaks begin.

9:00am-9:45am	Swim
9:45am-9:55am	Break
9:55am-10:40am	Swim
10:40am-10:50am	Break
10:50am-11:35am	Swim
11:35am-11:45am	Break

Table 1

12:30pm	Break begins
1:25pm	Break begins
2:20pm	Break begins
3:15pm	Break begins
4:10pm	Break begins
5:05pm	Break begins

Table 2

Since this question asks about what time the last break time begins, pay close attention to times at which breaks begin in table 1. The first break begins at 9:45am, second break at 10:40am, third break at 11:35am and so forth. These three break times suggest a pattern such that as the value of hours is increased by one: from 9_{am} to 10_{am} to 11_{am}, the value of minutes is decreased by five: from 45_{min} to 40_{min} to 35_{min}. Thus, the following breaks after 11:35am are 12:30pm, 1:25pm, 2:20 pm, and so on so forth as shown in table 2. Therefore, the last break time begins at 5:05pm.

PRACTICE TEST 4
MATHEMATICS PROBLEMS
50 Questions
Time — 60 minutes

Directions: Solve each problem and enter your answer by marking the circle on the answer sheet. Choose the best answer among the answer choices given.

1. $8 \times \frac{3}{4} - 6 \times \frac{2}{3} =$
   - A. 0
   - B. 1
   - C. 2
   - D. 3
   - E. 4

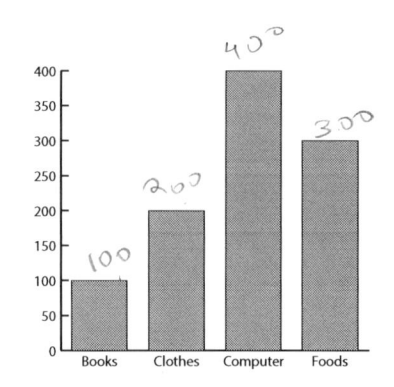

2. The graph shows the amount of money that Sue spent in November. Of the total amount of money spent, what percent did she spend on computer?
   - F. 40%
   - G. 35%
   - H. 30%
   - J. 25%
   - K. 20%

3. Joshua has seven marbles. If Jason has three less than twice the number of marbles that Joshua has, how many marbles does Jason have?
   - A. 11
   - B. 12
   - C. 13
   - D. 14
   - E. 15

4. What is the greatest prime factor of 52?
   - F. 2
   - G. 5
   - H. 7
   - J. 11
   - K. 13

5. What is the mean of the first five positive odd integers?
   - A. 4
   - B. 5
   - C. 6
   - D. 7
   - E. 8

6. Which of the following value is equal to $2(-1)^2 + 3(2)^2 + 4(-3)^2$?

   F. 184
   G. 125
   H. 73
   J. 50
   K. 38

7. If the length of a square is $3x + 2$, what is the perimeter of the square?

   A. $6x + 4$
   B. $6x + 8$
   C. $12x + 2$
   D. $12x + 8$
   E. $9x^2 + 4$

$$\frac{x-10}{4} + \frac{2x+40}{4} + \frac{x-30}{4}$$

8. If $x = 5$, evaluate the expression above.

   F. 3
   G. 5
   H. 7
   J. 9
   K. 10

9. If the volume of a pizza dough is doubled every two hours, in how many hours is the volume of the pizza dough eight times larger than the initial volume of the pizza dough?

   A. 6
   B. 7
   C. 8
   D. 9
   E. 10

10. In a set of 9, 4, x, 17, and 12, where $12 < x < 17$, which of the following is the median of the set?

    F. 17
    G. 12
    H. 9
    J. 4
    K. x

11. If $x + y = k$, what is $2x - y + x + 4y$ in terms of k ?

    A. k
    B. $1.5k$
    C. $2k$
    D. $2.5k$
    E. $3k$

12. If the ratio of a to b is $1 : 2$ and the ratio of b to c is $4 : 3$, which of the following is equal to the ratio of a to c ?

    F. $1 : 3$
    G. $1 : 2$
    H. $2 : 3$
    J. $2 : 5$
    K. $3 : 4$

13. If $\frac{2}{3}x + 3 = 5$, what is the value of $2x$?

    A. 3
    B. 5
    C. 6
    D. 7
    E. 9

14. If a positive integer is divided by 6, which of the following cannot be the remainder?

    F. 0
    G. 1
    H. 3
    J. 5
    K. 6

15. If $y = \frac{3x}{n}$, which of the following expression is equal to $4n$?

    A. $\frac{x}{12y}$
    B. $\frac{12x}{y}$
    C. $\frac{y}{3x+4}$
    D. $\frac{4y}{3x}$
    E. $\frac{y+4}{3x}$

16. Joshua has $900 in his savings account. If he saves $300 per week, in how many weeks will he have saved a total amount of $3000 in his savings account?

    F. 7
    G. 8
    H. 9
    J. 10
    K. 11

17. If Jason can type 150 words in 2 minutes, at this rate, how many words can he type in 3 minutes?

    A. 325
    B. 300
    C. 275
    D. 250
    E. 225

18. If $x - 2y = 6$ and $x + 2y = 10$, what is the value of $x + y$?

    F. 7
    G. 8
    H. 9
    J. 10
    K. 11

19. If a car travels at a rate of 42 miles per hour, how many miles does it travel in 50 minutes?

    A. 34
    B. 35
    C. 37
    D. 38
    E. 40

20. If the measures of interior angles of a triangles are $x - 5$, $x + 30$, and $2x + 15$, what is the measure of the largest angle?

    F. 105°
    G. 100°
    H. 90°
    J. 85°
    K. 75°

21. If the perimeter of a square is 16, what is the area of the square?

    A. 16
    B. 12
    C. 10
    D. 8
    E. 4

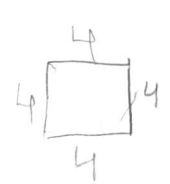

22. Let $x \spadesuit y = \frac{xy}{x+y}$. Which of the following is equal to $30 \spadesuit 6$?

    F. 1
    G. 2
    H. 3
    J. 4
    K. 5

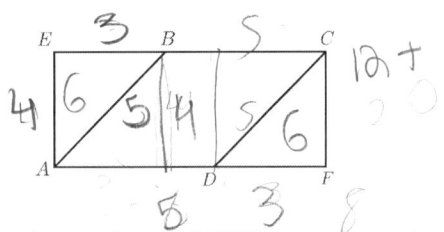

23. In the figure above, $AECF$ is a rectangle and $ABCD$ is a rhombus. If $BE = 3$ and $AE = 4$, what is the area of rectangle $AECF$?

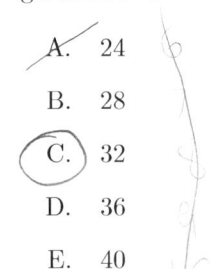

    A. 24
    B. 28
    C. 32
    D. 36
    E. 40

24. Two numbers, y and 3, are on the number line. If the midpoint of y and 3 is x, what is the value of y in terms of x?

    F. $2x + 3$
    G. $2x - 3$
    H. $3 - 2x$
    J. $6 + x$
    K. $6 - x$

25. Joshua starts walking due south 5 feet per second. At the same time, Jason starts walking due north 4 feet per second. If they start walking from the same location, how far apart are they after 9 seconds?

    A. 27 feet
    B. 36 feet
    C. 54 feet
    D. 72 feet
    E. 81 feet

26. In quadrilateral $ABCD$ shown above, $m\angle D = 90°$. If the ratio of the other three interior angles of the quadrilateral is $1 : 2 : 3$, what is the measure of the smallest interior angle of the quadrilateral in degrees?

    F. $45°$
    G. $60°$
    H. $90°$
    J. $120°$
    K. $135°$

27. If $x = 2$ is a solution to $3x - b = 5(x - 2)$, what is the value of b?

    A. 2
    B. 3
    C. 4
    D. 5
    E. 6

Yellow, Green, Blue, Red, Black, ···

28. In a clothing store, all the T-shirts are arranged in the pattern shown above. What is the color of the 37th T-shirt?

   F. Yellow
   G. Green
   H. Blue
   J. Red
   K. Black

29. What is the length of the diagonal of a square whose side length is 7?

   A. 14
   B. $7\sqrt{3}$
   C. $7\sqrt{2}$
   D. $5\sqrt{3}$
   E. $5\sqrt{2}$

30. If $\sqrt{x-5} = 2$, what is the value of x?

   F. 4
   G. 5
   H. 6
   J. 7
   K. 9

31. If $3^n = 2$, what is the value of 3^{n+1}?

   A. 3
   B. 4
   C. 6
   D. 7
   E. 9

32. $xy + y = x + 5$. If $x = 3$, what is the value of y?

   F. 5
   G. 4
   H. 3
   J. 2
   K. 1

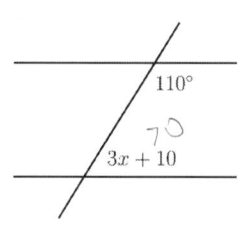

33. If the two lines are parallel, what is the value of $3x$?

   A. 60
   B. 50
   C. 40
   D. 30
   E. 20

34. Mr. Rhee reads 3 pages in 4 minutes. At the same rate, how many pages will he read in 2 hour?

   F. 30
   G. 45
   H. 60
   J. 75
   K. 90

35. If the length and width of a rectangle are $x - 2$ and $x + 2$, respectively, what is the area of the rectangle in terms of x?

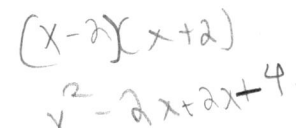

   A. $x^2 - 4x + 4$
   B. $x^2 - 4$
   C. x^2
   D. $4x - 8$
   E. $8x$

36. If February 4th is on Sunday, on what day of the week is February 24th?

   F. Thursday
   G. Friday
   H. Saturday
   J. Sunday
   K. Monday

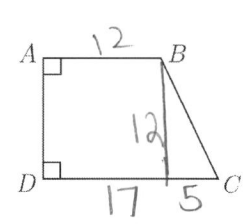

37. In the figure above, $AB = 12$, $CD = 17$, and $AD = 12$. What is the length of $\overline{BC}$?

   A. 11
   B. 12
   C. 13
   D. 14
   E. 15

38. Joshua starts running. For the first second, he is running due East 5 meters from the starting position. For the next second, he is running due West 2 meters. If he continues running in this pattern, how far is Joshua away from the starting position after 13 seconds?

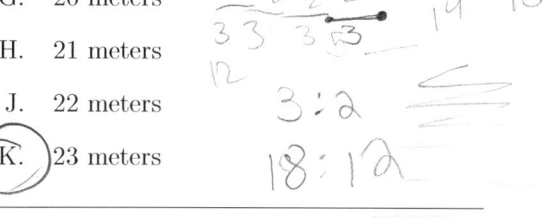

   F. 19 meters
   G. 20 meters
   H. 21 meters
   J. 22 meters
   K. 23 meters

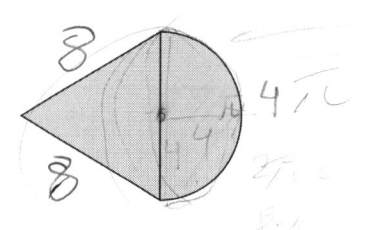

39. The figure above consists of an equilateral triangle and a semicircle. If the diameter of the semicircle is 8, what is the perimeter of the shaded area?

   A. $8 + 4\pi$
   B. $12 + 8\pi$
   C. $16 + 4\pi$
   D. $16 + 8\pi$
   E. $24 + 8\pi$

40. If you toss a coin twice, what is the probability that one head will be shown?

   F. $\frac{1}{4}$
   G. $\frac{1}{3}$
   H. $\frac{1}{2}$
   J. $\frac{3}{5}$
   K. $\frac{2}{3}$

$$ny - nx - 6n = 0$$

41. If $n > 0$, what is the value of $y - x$?

    A. 1
    B. 2
    C. 4
    D. 6
    E. 12

42. If $|x - 1| = 4$, what is the sum of the solutions?

    F. 5
    G. 4
    H. 3
    J. 2
    K. 1

43. There are three red marbles and a certain number of black and white marbles in a bag. If the probability of selecting a red marble is $\frac{1}{5}$, how many non-red marbles are in the bag?

    A. 13
    B. 12
    C. 11
    D. 10
    E. 9

Time	Average Speed
9 am to 10:30 am	40 miles per hour
10:30 am to 11:30 am	50 miles per hour
11:30 am to 12:30pm	Break
12:30 pm to 3:00 pm	60 miles per hour

44. According to the chart above, what is the total distance that Mr. Rhee traveled between 9 am to 3 pm?

    F. 220 miles
    G. 230 miles
    H. 240 miles
    J. 250 miles
    K. 260 miles

45. For all positive integers k, $k\blacklozenge$ is defined as the product of all positive consecutive even integers less than or equal to k. For instance, $6\blacklozenge = 6 \times 4 \times 2$. What is the value of $\frac{16\blacklozenge}{15\blacklozenge}$?

    A. 1
    B. 14
    C. 16
    D. 20
    E. 26

46. If the mean of two consecutive even integers is 23, what is the value of the smaller integer?

    F. 26
    G. 24
    H. 23
    J. 22
    K. 20

47. Set S consists of 5 positive integers: 6, x, 10, y, and 15. If the mode and the mean of set S are 10 and 11 respectively, what is the product of x and y?

   A. 150
   B. 140
   C. 130
   D. 120
   E. 110

48. There are thirty questions on a math exam worth a total of one hundred points. Each question is worth either three points or five points. How many questions on the math exam are worth five points?

   F. 4
   G. 5
   H. 6
   J. 7
   K. 8

49. Eight cards, each labeled with a number from 1 through 8, are in a bag. What is the smallest number of cards you need to select so that at least one prime number is guaranteed among the selected cards?

   A. 5
   B. 4
   C. 3
   D. 2
   E. 1

50. A penny, a nickel, a dime, and a quarter are in a bag. If you select two coins at random, how many different total values are possible?

   F. 5
   G. 6
   H. 7
   J. 8
   K. 9

STOP

SOLOMON ACADEMY   Distribution or replication of any part of this page is prohibited.   **TEST 4 SOLUTIONS**

Answers and Solutions
Practice Test 4

Answers

1. C	2. F	3. A	4. K	5. B
6. J	7. D	8. G	9. A	10. G
11. E	12. H	13. C	14. K	15. B
16. F	17. E	18. H	19. B	20. J
21. A	22. K	23. C	24. G	25. E
26. F	27. E	28. G	29. C	30. K
31. C	32. J	33. A	34. K	35. B
36. H	37. C	38. K	39. C	40. H
41. D	42. J	43. B	44. K	45. C
46. J	47. B	48. G	49. A	50. G

Solutions

1. (C)
$$\left(8 \times \frac{3}{4}\right) - \left(6 \times \frac{2}{3}\right) = 6 - 4 = 2$$

2. (F)

    The total amount of money that Sue spent in November is $100 + $200 + $400 + $300 = $1000. Since Sue spent $400 on a computer, the percent that she spent on a computer out of the total amount of money is $\frac{\$400}{\$1000} = 0.4 = 40\%$.

3. (A)

    Joshua has 7 marbles. Since Jason has three less than twice the number of marbles that Joshua has, Jason has $2(7) - 3 = 11$ marbles.

4. (K)

    The factors of 52 are 1, 2, 4, 13, 26, and 52. Among the six factors, the greatest prime factor is 13.

5. (B)

    The sum of the first five positive odd integers is $1 + 3 + 5 + 7 + 9 = 25$. Therefore, the mean of the first five positive odd integers is $\frac{25}{5} = 5$.

www.solomonacademy.net

SOLOMON ACADEMY — TEST 4 SOLUTIONS

6. (J)
$$2(-1)^2 + 3(2)^2 + 4(-3)^2 = 2(1) + 3(4) + 4(9)$$
$$= 50$$

7. (D)

> Tips: The perimeter of a square is four times the length of the side of the square.

Since the length of the square is $3x + 2$, the perimeter of the square is $4(3x + 2) = 12x + 8$.

8. (G)

> Tips: Simplify the expression first before substituting 5 for x in the expression.

$$\frac{x-10}{4} + \frac{2x+40}{4} + \frac{x-30}{4} = \frac{x-10+2x+40+x-30}{4}$$
$$= \frac{4x}{4}$$
$$= x$$

Since the expression simplifies to x, the value of expression $\frac{x-10}{4} + \frac{2x+40}{4} + \frac{x-30}{4}$ is 5.

9. (A)

Let V_0 be the initial volume of the pizza dough. Since the volume of the pizza dough is doubled every two hours,

$$\text{In 2 hours} = 2V_0$$
$$\text{In 4 hours} = 2(2V_0) = 4V_0$$
$$\text{In 6 hours} = 2(4V_0) = 8V_0$$

Therefore, in 6 hours, the volume of the pizza dough is eight times larger than the initial volume of the pizza dough.

10. (G)

Since $12 < x < 17$, arrange the numbers in the set from least to greatest: 4, 9, 12, x, and 17. There are five numbers in the set. Thus, the median is the third number in the set. Therefore, x is the median of the set.

11. (E)

$$2x - y + x + 4y = 2x + x - y + 4y \qquad \text{Simplify}$$
$$= 3x + 3y \qquad \text{Factor out 3}$$
$$= 3(x + y) \qquad \text{Substitute } k \text{ for } x + y$$
$$= 3k$$

Therefore, $2x - y + x + 4y$ in terms of k is $3k$.

SOLOMON ACADEMY

Distribution or replication of any part of this page is prohibited.

TEST 4 SOLUTIONS

12. (H)

Since $\frac{a}{b} = \frac{1}{2}$ and $\frac{b}{c} = \frac{4}{3}$, the ratio $\frac{a}{c}$ can be calculated using the two given ratios.

$$\frac{a}{c} = \frac{a}{b} \times \frac{b}{c}$$
$$= \frac{1}{2} \times \frac{4}{3}$$
$$= \frac{2}{3}$$

Therefore, the ratio of a to c is $2 : 3$.

13. (C)

$$\frac{2}{3}x + 3 = 5 \qquad \text{Subtract 3 from each side}$$
$$\frac{2}{3}x = 2 \qquad \text{Multiply each side by 3}$$
$$2x = 6$$

Therefore, the value of $2x$ is 6.

14. (K)

> **Tips** When a positive integer is divided by n, the possible values of the remainder are $0, 1, 2, \cdots, n-1$.

Since the positive integer is divided by 6, the possible values of the remainder are 0, 1, 2, 3, 4, and 5. Therefore, (K) is the correct answer.

15. (B)

> **Tips** If $a = \frac{b}{c}$, then $c = \frac{b}{a}$.

Since $y = \frac{3x}{n}$, $n = \frac{3x}{y}$. Therefore, $4n = 4 \times \frac{3x}{y} = \frac{12x}{y}$.

16. (F)

Let x be the number of weeks that Joshua needs to save the total amount of $3000. Since Joshua will save $300 per week, the amount that Joshua will save in x weeks will be $300x$. Thus,

$$300x + 900 = 3000 \qquad \text{Subtract 900 from each side}$$
$$300x = 2100 \qquad \text{Divide each side by 300}$$
$$x = 7$$

Therefore, Joshua needs 7 weeks to save the total amount of $3000.

17. (E)

If Jason can type 150 words in 2 minutes, he can type $150 \div 2 = 75$ words in 1 minute. Therefore, he can type $75 \times 3 = 225$ words in 3 minutes.

18. (H)

Add two equations and solve for x.

$$x - 2y = 6$$
$$x + 2y = 10 \quad \text{Add two equations}$$
$$2x = 16 \quad \text{Divide each side by 2}$$
$$x = 8$$

Substitute 8 for x in the first equation $x - 2y = 6$ and solve for y.

$$x - 2y = 6 \quad \text{Substitute 8 for } x$$
$$8 - 2y = 6 \quad \text{Subtract 8 from each side}$$
$$-2y = -2 \quad \text{Divide each side by } -2$$
$$y = 1$$

Thus, $x = 8$ and $y = 1$. Therefore, the value of $x + y = 8 + 1 = 9$.

19. (B)

42 miles per hour means that the car travels 42 miles in one hour. There are 60 minutes in one hour. Set up a proportion in terms of miles and minutes.

$$42_{\text{miles}} : 60_{\text{minutes}} = x_{\text{miles}} : 50_{\text{minutes}}$$
$$\frac{42}{60} = \frac{x}{50} \quad \text{Use cross product property}$$
$$60x = 42 \times 50$$
$$x = 35$$

Therefore, the car travels 35 miles in 50 minutes.

20. (J)

Tips The sum of the measures of interior angles of a triangle is 180°.

Since the sum of the measures of interior angles of a triangle is 180°,

$$x - 5 + x + 30 + 2x + 15 = 180$$
$$4x + 40 = 180 \quad \text{Subtract 40 from each side}$$
$$4x = 140 \quad \text{Divide each side by 4}$$
$$x = 35$$

Therefore, the measure of the largest angle is $2x + 15 = 2(35) + 15 = 85°$.

21. (A)

The perimeter of the square is 16. This means that the length of the side of the square is 4. Therefore, the area of the square is $4^2 = 16$.

22. (K)

Since $x \spadesuit y = \frac{xy}{x+y}$, $30 \spadesuit 6 = \frac{30 \times 6}{30+6} = \frac{180}{36} = 5$.

SOLOMON ACADEMY — TEST 4 SOLUTIONS

Distribution or replication of any part of this page is prohibited.

23. (C)

    $AECF$ is a rectangle so $m\angle E = 90°$. $\triangle AEB$ is a right triangle with $BE = 3$ and $AE = 4$. Use the Pythagorean theorem to find the length of $\overline{AB}$: $AB^2 = 3^2 + 4^2$. Thus, $AB = 5$. $ABCD$ is a rhombus. This means that $AB = BC = 5$. Thus, $EC = BE + BC = 3 + 5 = 8$. Therefore, the area of rectangle $AECF$ is $AE \times EC = 4 \times 8 = 32$.

24. (G)

    > Tips: If m is the midpoint of p and q on the number line, $m = \frac{p+q}{2}$ or $2m = p + q$.

    x is the midpoint of y and 3 on the number line. Thus, $x = \frac{y+3}{2}$ or $2x = y + 3$, which implies that $y = 2x - 3$. Therefore, y in terms of x is $2x - 3$.

25. (E)

    Joshua is walking due south 5 feet per second and Jason is walking due north 4 feet per second. They are walking opposite direction and are 9 feet apart every second. Therefore, they are $9 \times 9 = 81$ feet apart in 9 seconds.

26. (F)

    In quadrilateral $ABCD$, $m\angle D = 90°$. Since the ratio of the other three interior angles of the quadrilateral is $1:2:3$, let x, $2x$, and $3x$ be the measure of angle A, B, and C, respectively. The sum of the measures of the interior angles of a quadrilateral is $360°$. Thus,

    $$x + 2x + 3x + 90 = 360$$
    $$6x = 270$$
    $$x = 45$$

    Therefore, the measure of the smallest interior angle of the quadrilateral is $x = 45°$.

27. (E)

    Substitute 2 for x in the equation $3x - b = 5(x - 2)$ and solve for b.

    $$3x - b = 5(x - 2) \qquad \text{Substitute 2 for } x$$
    $$6 - b = 5(2 - 2)$$
    $$6 - b = 0$$
    $$b = 6$$

    Therefore, the value of b is 6.

28. (G)

    The pattern consists of five different colors: Yellow, Green, Blue, Red, and Black and it is repeated. This means that every fifth shirt is Black. Thus, the $10^{\text{th}}, 15^{\text{th}}, \cdots, 30^{\text{th}}$, and 35th shirts are Black, the 36th shirt is Yellow, and the 37th shirt is Green.

29. (C)

If the square is divided diagonally as shown below, it forms two 45°-45°-90° special right triangles whose sides are in the ratio $1 : 1 : \sqrt{2}$.

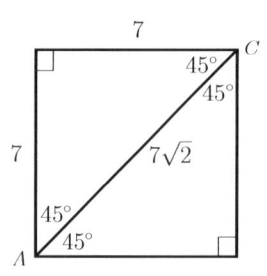

The length of the diagonal of the square is the same as the length of the hypotenuse of the two right triangles, AC. Since the length of the hypotenuse is $\sqrt{2}$ times the length of each leg, $AC = 7\sqrt{2}$. Therefore, the length of the diagonal of the square is $7\sqrt{2}$.

30. (K)

$$\sqrt{x-5} = 2 \quad \text{Square each side}$$
$$x - 5 = 4 \quad \text{Add 5 to each side}$$
$$x = 9$$

Therefore, the value of x is 9.

31. (C)

> Tips: Use the property of exponents: $a^{x+y} = a^x \cdot a^y$

Based on the property of exponents, $3^{n+1} = 3^n \times 3^1$. Thus,

$$3^{n+1} = 3^n \times 3^1 \quad \text{Since } 3^n = 2$$
$$= 2 \times 3$$
$$= 6$$

Therefore, the value of 3^{n+1} is 6.

32. (J)

Substitute 3 for x in the equation $xy + y = x + 5$ and solve for y.

$$xy + y = x + 5 \quad \text{Substitute 3 for } x$$
$$3y + y = 3 + 5 \quad \text{Simplify}$$
$$4y = 8 \quad \text{Solve for } y$$
$$y = 2$$

Therefore, the value of y is 2.

SOLOMON ACADEMY

Distribution or replication of any part of this page is prohibited.

TEST 4 SOLUTIONS

---

33. (A)

    Since the two lines are parallel, angles $3x + 10$ and $110°$ are consecutive angles and supplementary angles whose sum of their measures is $180°$. Thus,

    $$3x + 10 + 110 = 180$$
    $$3x = 60$$
    $$x = 20$$

    Therefore, the value of $3x = 3(20) = 60$.

34. (K)

    There are 60 minutes in one hour. Thus, there are 120 minutes in 2 hours. Set up a proportion in terms of pages and minutes.

    $$3_{\text{pages}} : 4_{\text{minutes}} = x_{\text{pages}} : 120_{\text{minutes}}$$
    $$\frac{3}{4} = \frac{x}{120} \qquad \text{Use cross product property}$$
    $$4x = 360$$
    $$x = 90$$

    Therefore, Mr. Rhee will read 90 pages in 2 hours.

35. (B)

    Tips
    1. The area of a rectangle is $A = lw$, where l is the length and w is the width.
    2. Difference of squares formula: $(x+y)(x-y) = x^2 - y^2$

    The length and width of the rectangle are $x - 2$ and $x + 2$.

    $$\text{Area of the rectangle} = lw$$
    $$= (x+2)(x-2)$$
    $$= x^2 - 2^2$$
    $$= x^2 - 4$$

    Therefore, the area of the rectangle in terms of x is $x^2 - 4$.

36. (H)

    The days of the week repeat every 7 days. If February 4th falls on a Sunday, the 11th, 18th, and 25th are Sundays. Since February 24th is the day before February 25th, it is a Saturday.

SOLOMON ACADEMY
Distribution or replication of any part of this page is prohibited.

TEST 4 SOLUTIONS

37. (C)

> **Tips**: If a triangle is a right triangle, use the Pythagorean theorem: $c^2 = a^2 + b^2$, where c is the hypotenuse, a and b are the legs of the right triangle.

Segment BE is drawn from vertex B to $\overline{DC}$ such that $\overline{BE} \parallel \overline{AD}$ as shown in the figure below. This means that $\overline{BE} \perp \overline{EC}$.

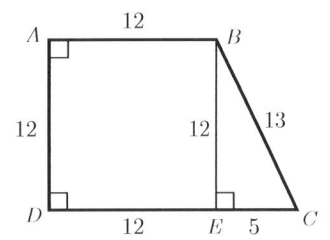

$ABED$ is a square with side length of 12. Thus, $AB = BE = DE = 12$. Since $CD = 17$, $CE = 5$. $\triangle CBE$ is a right triangle with $BE = 12$ and $CE = 5$. To find the length of $\overline{BC}$, use the Pythagorean theorem: $c^2 = 5^2 + 12^2$, where c is the length of $\overline{BC}$. Therefore, the length of $\overline{BC}$ is 13.

38. (K)

Joshua starts running. For the first second, he is running due East 5 meters from the starting position. For the next second, he is running due West 2 meters. This means that Joshua is 3 meters farther away from the starting position every 2 seconds. After 12 seconds, Joshua is $6 \times 3 = 18$ meters away from the starting position. At the 13th second, Joshua is running another 5 meters away from the starting position. Therefore, after 13 seconds, Joshua is $18 + 5 = 23$ meters away from the starting position.

39. (C)

In the figure below, the diameter of the semicircle is 8. This means that the length of the side of the equilateral triangle is 8 and the radius of the semicircle is 4.

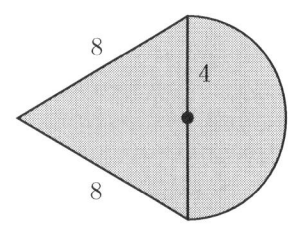

The perimeter is the distance around the figure. Since the diameter of the semicircle is inside the shaded region, it should be excluded from the perimeter of the shaded region. Thus, the perimeter of the shaded region equals the sum of the lengths of two sides of the equilateral triangle and half the circumference of the circle with a radius of 4. Therefore, the perimeter of the shaded region is $8 + 8 + \frac{1}{2}(2\pi(4)) = 16 + 4\pi$.

SOLOMON ACADEMY — TEST 4 SOLUTIONS

40. (H)

When a coin is tossed, there are two possible outcomes: H(head) or T(tail). If you toss a coin twice, there are four possible outcomes: $H\ H$, $H\ T$, $T\ H$, and $T\ T$. Out of 4 possible outcomes, there are 2 outcomes that have one head: $H\ T$ and $T\ H$. Therefore, the probability that one head will be shown is $\frac{2}{4} = \frac{1}{2}$.

41. (D)

Since $n > 0$, divide each side of the equation $ny - nx - 6n = 0$ by n and solve for $y - x$.

$$\frac{ny - nx - 6n}{n} = \frac{0}{n} \quad \text{Divide each side by } n$$

$$\frac{ny}{n} - \frac{nx}{n} - \frac{6n}{n} = 0 \quad \text{Simplify}$$

$$y - x - 6 = 0 \quad \text{Add 6 to each side}$$

$$y - x = 6$$

Therefore, the value of $y - x$ is 6.

42. (J)

$$|x - 1| = 4$$
$$x - 1 = \pm 4$$

$x - 1 = 4$ or $x - 1 = -4$
$x = 5$ or $x = -3$

Therefore, the sum of the solutions is $5 + (-3) = 2$.

43. (B)

Let x be the total number of marbles in the bag. There are three red marbles and the probability of selecting a red marble is $\frac{1}{5}$.

$$\text{Probability of selecting a red marble} = \frac{\text{Number of red marbles}}{\text{Total number of marbles}}$$

$$\frac{1}{5} = \frac{3}{x} \quad \text{Cross multiply}$$

$$x = 15$$

Thus, the total number of marbles in the bag is 15. Therefore, there are $15 - 3 = 12$ non-red marbles in the bag.

SOLOMON ACADEMY — TEST 4 SOLUTIONS

44. (K)

> Tips: Use the distance formula: distance = speed × time.

Below shows the distance that Mr. Rhee traveled during the different time intervals.

Time	Average Speed	Distance traveled
9 am to 10:30 am	40 miles per hour	40 mph × 1.5 hour = 60 miles
10:30 am to 11:30 am	50 miles per hour	50 mph × 1 hour = 50 miles
11:30 am to 12:30pm	Break	0 mile
12:30 pm to 3:00 pm	60 miles per hour	60 mph × 2.5 hour = 150 miles

Therefore, the total distance that Mr. Rhee traveled between 9 am to 3 pm is $60+50+150 = 260$ miles.

45. (C)

$k\blacklozenge$ is defined as the product of all positive consecutive even integers less than or equal to k.

$$\frac{16\blacklozenge}{15\blacklozenge} = \frac{16 \times 14 \times 12 \times 10 \times 8 \times 6 \times 4 \times 2}{14 \times 12 \times 10 \times 8 \times 6 \times 4 \times 2}$$
$$= 16$$

Therefore, the value of $\frac{16\blacklozenge}{15\blacklozenge}$ is 16.

46. (J)

> Tips: Sum of two integers = mean of the two integers × 2

Let x be the smaller consecutive even integer. Then, the larger consecutive even integer can be expressed as $x+2$ and the sum of the two integers can be expressed as $x+x+2$. Since the mean of the two integers is 23, the sum of the integers is $2 \times 23 = 46$. Thus,

$$x + x + 2 = 46 \qquad \text{Subtract 2 from each side}$$
$$2x = 44 \qquad \text{Divide each side by 2}$$
$$x = 22$$

Therefore, the value of the smaller integer is 22.

47. (B)

Set S consists of 5 positive integers: 6, x, 10, y, and 15. The mean of the five integers in set S is 11, which means that the sum of the five integers is $5 \times 11 = 55$. Set up an equation in terms of the sum and find the value of $x+y$.

$$x + y + 6 + 10 + 15 = 55 \qquad \text{Subtract 31 from each side}$$
$$x + y = 24$$

Thus, $x+y = 24$. Since the mode of set S is 10, either x or y is 10. If $x = 10$, $y = 14$ so that $x+y = 24$. For the same reason, if $y = 10$, $x = 14$. Therefore, the product of xy is $xy = 10(14) = 140$.

SOLOMON ACADEMY — TEST 4 SOLUTIONS

48. (G)

There are two different type of questions on the math exam. One is worth 3 points and another is worth 5 points. Let x be the number of questions worth 5 points. Since there are 30 questions on the math exam, $30 - x$ is the number of questions worth 3 points. Below shows how to obtain the sum of points for each type of questions.

	A question worth 5 points	A question worth 3 points	Total
Number of questions	x	$30 - x$	30
Sum of points	$5x$	$3(30 - x)$	100

Since the math exam is worth a total of 100 points, set up an equation in terms of the sum of points shown on the table above.

$$5x + 3(30 - x) = 100 \quad \text{Expand } 3(30 - x)$$
$$5x - 3x + 90 = 100 \quad \text{Subtract 90 from each side}$$
$$2x = 10 \quad \text{Divide each side by 2}$$
$$x = 5$$

Therefore, the number of questions that are worth 5 points is 5.

49. (A)

There are four non-prime numbers from 1 to 8: 1, 4, 6, and 8. There are four prime numbers: 2, 3, 5, and 7. If all four non-prime numbers are selected, the next number will be a prime number. Therefore, five cards must be selected to guarantee that at least one prime number is selected.

50. (G)

A penny, a nickel, a dime, and a quarter are in a bag. Two coins are selected at random. The table below shows different possible total values if two coins are selected. Define P, N, D, and Q as a penny, a nickel, a dime, and a quarter, respectively.

$P + N$	6 cents
$P + D$	11 cents
$P + Q$	26 cents
$N + D$	15 cents
$N + Q$	30 cents
$D + Q$	35 cents

Therefore, there are six different possible total values.

SOLOMON ACADEMY

PRACTICE TEST 5

PRACTICE TEST 5
MATHEMATICS PROBLEMS
50 Questions
Time — 60 minutes

Directions: Solve each problem and enter your answer by marking the circle on the answer sheet. Choose the best answer among the answer choices given.

1. $2\frac{3}{4} \div 3\frac{2}{3} =$

    A. $\frac{4}{5}$
    B. $\frac{3}{4}$
    C. $\frac{2}{3}$
    D. $\frac{1}{2}$
    E. $\frac{1}{3}$

2. $100 - 99 + 99 - 98 + 98 - 97 =$

    F. 3
    G. 4
    H. 5
    J. 6
    K. 7

3. What is the average of $x + 4$ and $3x + 8$?

    A. $3x^2 + 32$
    B. $4x + 12$
    C. $4x + 6$
    D. $2x + 12$
    E. $2x + 6$

4. What is the sum of the factors of 16?

    F. 24
    G. 27
    H. 30
    J. 31
    K. 35

5. Which of the following expression is equal to $\dfrac{x^2 \cdot x^2 \cdot x^2}{x^3}$?

    A. x^2
    B. x^3
    C. x^4
    D. x^5
    E. x^6

6. Joshua is x years old now. If he will be 21 years old in $x + 3$ years, how old is Joshua now?

    F. 8
    G. 9
    H. 10
    J. 11
    K. 12

147   www.solomonacademy.net

7. If $\frac{x}{2} + \frac{x}{3} = 5$, what is the value of x?

   A. 12
   B. 10
   C. 8
   D. 6
   E. 3

8. Which of the following integer is the smallest prime number?

   F. 0
   G. 1
   H. 2
   J. 3
   K. 5

9. If the product of 3 and a number equals the sum of 3 and the number, what is the number?

   A. 2
   B. 1
   C. 0
   D. $\frac{1}{2}$
   E. $\frac{3}{2}$

10. Evaluate $2|x-5| - 3|1-x|$ when $x = 2$.

    F. 1
    G. 2
    H. 3
    J. 4
    K. 5

11. If a positive integer is divided by either 3 or 4, the remainders are both 1. What is the positive integer?

    A. 25
    B. 24
    C. 18
    D. 15
    E. 14

12. If the circumference and the area of a circle are the same, what is the radius of the circle?

    F. 1
    G. 2
    H. 3
    J. 4
    K. 5

13. If $-3 < x \leq 2$, what is the sum of all possible integer values of x?

    A. 0
    B. 1
    C. 2
    D. 3
    E. 4

14. If $\frac{y}{x} = \frac{1}{2}$, what is the value of $\frac{x^2}{y^2}$?

    F. $\frac{1}{4}$
    G. $\frac{1}{2}$
    H. 1
    J. 2
    K. 4

15. If $3(x+y) = 12$, what is x in terms of y?

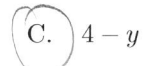

    A. $12 + 3y$
    B. $4 + 3y$
    C. $4 - y$
    D. $4 - 3y$
    E. $12 - 3y$

16. How many positive integers are there between 24 and 81 inclusive?

    F. 47
    G. 43
    H. 57
    J. 58
    K. 59

17. Joshua has half as many coins as Jason. Mr. Rhee has twice as many coins as Jason. If Jason and Mr. Rhee have 60 coins together, how many coins does Joshua have?

    A. 6
    B. 7
    C. 8
    D. 9
    E. 10

18. If the price of a pencil increases from $1.20 to $1.68, what is the percent increase?

    F. 25 percent
    G. 30 percent
    H. 32 percent
    J. 35 percent
    K. 40 percent

19. If x is an integer that satisfies $x^3 + 1 = 65$, what is the value of $x^2 + 1$?

    A. 5
    B. 10
    C. 17
    D. 26
    E. 37

20. What is the median of the set of all factors of 36?

    F. 9
    G. 8
    H. 7.5
    J. 6
    K. 5

21. In the figure above, 4 congruent squares are placed side by side to form a rectangle. If the area of the rectangle is 36, what is the perimeter of the rectangle?

    A. 45
    B. 40
    C. 35
    D. 30
    E. 25

22. The mean of four numbers is 10. If 10 is added, what is the mean of the five numbers?

F. 10
G. 9
H. 8
J. 7
K. 6

23. If Mr. Rhee traveled 40 miles in 40 minutes, how fast, in miles per hour, did he travel?

A. 64 miles per hour
B. 60 miles per hour
C. 56 miles per hour
D. 50 miles per hour
E. 40 miles per hour

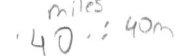

24. In the equation $xy + xz = 35$, what is the value of x if $y = 3$ and $z = 4$?

F. 2
G. 3
H. 4
J. 5
K. 6

25. Which of the following expression must be equal to $(x - 2)(x - 3)$?

A. $2x - 5$
B. $x^2 + 6$
C. $x^2 - 5x + 6$
D. $x^2 + 5x + 6$
E. $x^2 - 6x - 5$

26. One-third of the students in a class are taking science classes. One half of the students taking science classes are taking chemistry class. What fractional part of the students in the class are taking chemistry class?

F. $\frac{1}{6}$
G. $\frac{1}{5}$
H. $\frac{1}{2}$
J. $\frac{2}{3}$
K. $\frac{5}{6}$

27. The lengths of two segments $\overline{BA}$ and $\overline{CD}$ are 7 and 8 respectively. The two segments $\overline{BA}$ and $\overline{CD}$ are overlapped so that the four points A, C, B, and D are arranged in that order and they become collinear points. If $BC = 6$, what is the length of $\overline{AD}$?

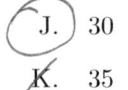

A. 13
B. 12
C. 11
D. 10
E. 9

28. There are forty students in a class. They are taking either math or chemistry or both. The number of students taking math is twice as many as the number of students taking chemistry. If there are five students taking both math and chemistry, what is the number of students taking math?

F. 15
G. 20
H. 25
J. 30
K. 35

29. If $3 \times 3^2 \times 3^3 = 3^n$, what is the value of n?

    A. 2
    B. 3
    C. 4
    D. 5
    E. 6

30. The volume of a cylinder is the base area times the height. If the volume of the cylinder is 54π and the height is 9, what is the base area of the cylinder?

    F. 6π
    G. 8π
    H. 10π
    J. 12π
    K. 16π

31. If $2x + 3y = 5$, which of the following is equal to $3(4x + 6y)$?

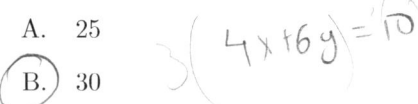

    A. 25
    B. 30
    C. 35
    D. 40
    E. 45

32. What percent of 15 is 12?

    F. $33\frac{1}{3}\%$
    G. $66\frac{2}{3}\%$
    H. 80%
    J. 85%
    K. 90%

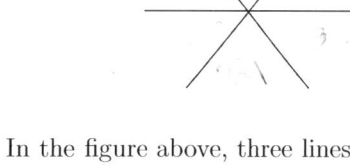

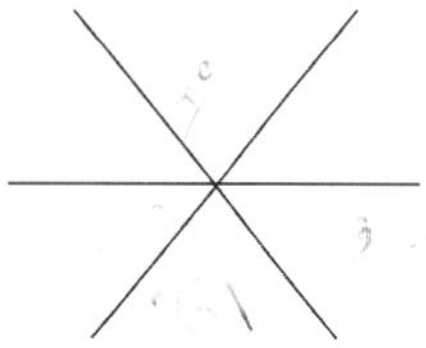

33. In the figure above, three lines intersect at one point to form six angles. If the degree measures of the six angles are integer values and they are not all congruent, what is the smallest possible degree measure of the largest angle?

    A. 59
    B. 60
    C. 61
    D. 177
    E. 178

34. In order to make a two-digit number, two numbers are selected at random from digits 1 through 4 without replacement. How many two-digit numbers are possible?

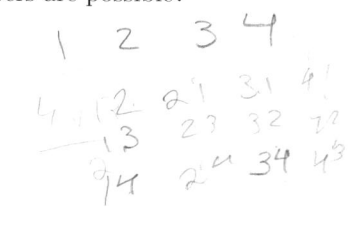

    F. 6
    G. 9
    H. 12
    J. 15
    K. 18

x	2	3	4	5
y	3	7	11	k

35. If the pattern shown on the table above continues, what is the value of k?

    A. 19
    B. 18
    C. 17
    D. 16
    E. 15

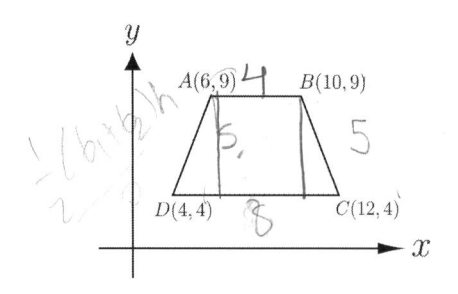

36. In the figure above, the quadrilateral $ABCD$ is a trapezoid. What is the area of the trapezoid?

F. 30
G. 40
H. 50
J. 60
K. 70

37. A print shop has two machines: machine A and machine B. Machine A can print 12 posters in 10 minutes and machine B can print 17 posters in 15 minutes. In the first half hour, only machine A prints posters. In the next hour, machine A and B print together. What is the total number of posters that machine A and machine B print?

A. 126
B. 128
C. 140
D. 176
E. 208

38. If the measures of the interior angles of a triangle are $\frac{x}{2}$, $\frac{x}{6}$, and x, what is the measure of the smallest angle?

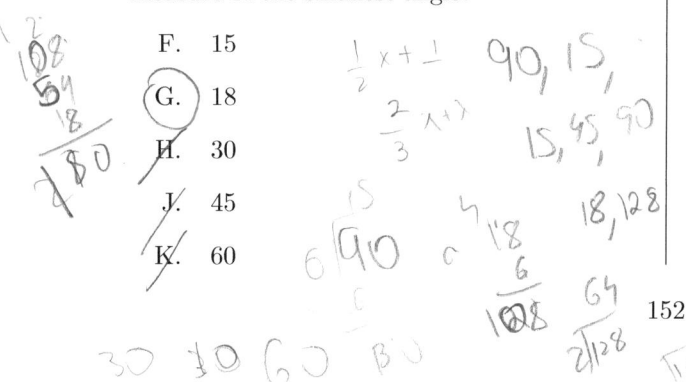

F. 15
G. 18
H. 30
J. 45
K. 60

x	2	3	5
y	-1	2	8

39. The table above shows three ordered pairs on a line. What is the slope of the line?

A. 3
B. 2
C. 1
D. 0
E. -2

40. If both n and k are prime numbers, which of the following statement must be true?

F. nk is an even number.
G. greatest common factor of n and k is 1.
H. $n + k$ is an even number.
J. $n + k$ is divisible by 2.
K. $n + k$ is a prime number.

41. Which of the following expressions is equal to $8(4)^3$?

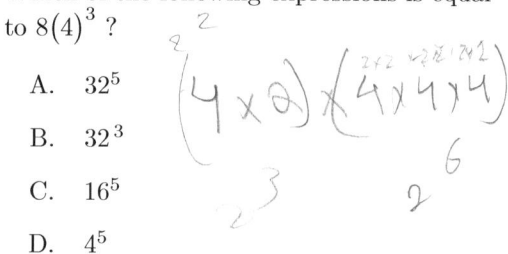

A. 32^5
B. 32^3
C. 16^5
D. 4^5
E. 2^9

42. If $x + y = 10$ and $x - y = 4$, what is the value of $x^2 - y^2$?

   F. 58
   G. 40
   H. 35
   J. 24
   K. 16

43. Mclean is 16 miles due north of Centreville. Brambleton is 12 miles due west of Centreville on a map. How many miles are Mclean and Brambleton apart?

   A. 18
   B. 19
   C. 20
   D. 26
   E. 28

44. It takes one minute to fill $\frac{2}{5}$ of a water tank. How many more minutes will it take to fill the remaining water tank?

   F. $\frac{6}{5}$ minutes
   G. $\frac{5}{4}$ minutes
   H. $\frac{5}{3}$ minute
   J. $\frac{4}{3}$ minute
   K. $\frac{3}{2}$ minute

45. Two standard dice are rolled. What is the probability that the sum of the two numbers on top of each die is either 8 or 10?

   A. $\frac{2}{9}$
   B. $\frac{1}{4}$
   C. $\frac{1}{3}$
   D. $\frac{5}{12}$
   E. $\frac{1}{2}$

46. A number is decreased by 3 and then the result is doubled. If the new result is six more than half the original number, what is the value of the original number?

   F. 14
   G. 12
   H. 10
   J. 8
   K. 6

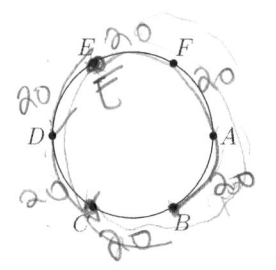

47. In the circular track whose circumference is 120 meters shown above, Joshua and Jason are running in opposite directions from the starting position, A. Joshua is running at 4 meters per second clockwise and Jason is running at 2 meters per second counterclockwise. Which of the following lettered position represents the second time Joshua and Jason meet after they begin running? (Assume that all the lettered positions are equally spaced.)

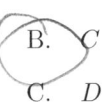

   A. B
   B. C
   C. D
   D. E
   E. F

48. If two lines $ax + 3y = 4$ and $y = \frac{1}{2}x$ are parallel, what is the value of a?

   F. $\frac{3}{2}$
   G. $\frac{4}{3}$
   H. $\frac{3}{4}$
   J. $-\frac{2}{3}$
   K. $-\frac{3}{2}$

49. If the mean and the median of a set of five distinct positive integers is 8 and 5 respectively, what is the greatest possible value of the largest integer in the set?

   A. 30
   B. 32
   C. 34
   D. 36
   E. 38

50. The length of cube A is 8 inches. If the length of cube A is increased by 25% to form cube B, how much greater, in square inches, is the surface area of cube B than the surface area of cube A?

   F. 132 square inches
   G. 164 square inches
   H. 192 square inches
   J. 216 square inches
   K. 256 square inches

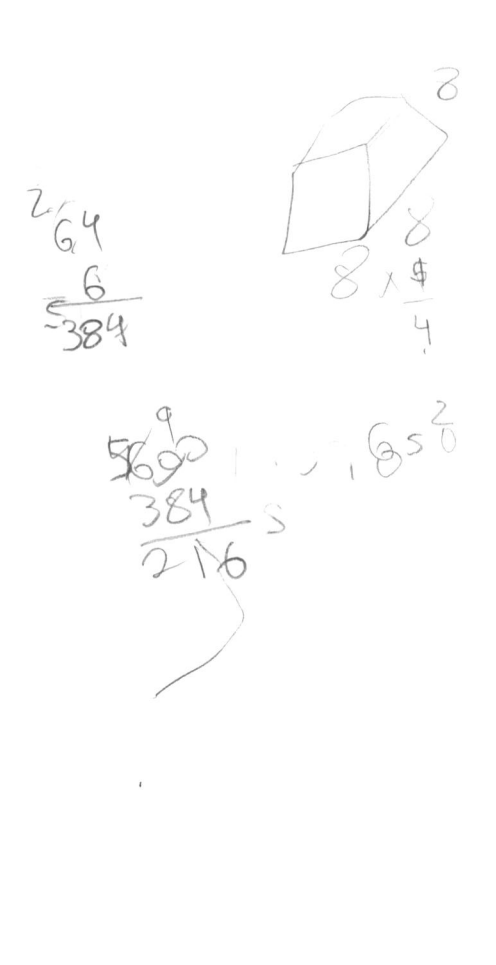

STOP

SOLOMON ACADEMY — TEST 5 SOLUTIONS

Answers and Solutions
Practice Test 5

Answers

1. B	2. F	3. E	4. J	5. B
6. G	7. D	8. H	9. E	10. H
11. A	12. G	13. A	14. K	15. C
16. J	17. E	18. K	19. C	20. J
21. D	22. F	23. B	24. J	25. C
26. F	27. E	28. J	29. E	30. F
31. B	32. H	33. C	34. H	35. E
36. F	37. D	38. G	39. A	40. G
41. E	42. G	43. C	44. K	45. A
46. J	47. B	48. K	49. D	50. J

Solutions

1. (B)
$$2\frac{3}{4} \div 3\frac{2}{3} = \frac{11}{4} \div \frac{11}{3}$$
$$= \frac{11}{4} \times \frac{3}{11}$$
$$= \frac{3}{4}$$

2. (F)
$$100 + (-99 + 99) + (-98 + 98) - 97 = 100 - 97$$
$$= 3$$

3. (E)
$$\text{Average} = \frac{\text{Sum}}{2}$$
$$= \frac{x + 4 + 3x + 8}{2}$$
$$= \frac{4x + 12}{2}$$
$$= 2x + 6$$

4. (J)

   The factors of 16 are 1, 2, 4, 8, and 16. Therefore, the sum of all the factors of 16 is $1+2+4+8+16 = 31$.

5. (B)

   Tips Use the properties of exponents.
   1. $a^x \cdot a^y = a^{x+y}$
   2. $\frac{a^x}{a^y} = a^{x-y}$

   $$\frac{x^2 \cdot x^2 \cdot x^2}{x^3} = \frac{x^{2+2+2}}{x^3}$$
   $$= \frac{x^6}{x^3}$$
   $$= x^3$$

6. (G)

   Joshua is x years old. In $x + 3$ years, he will be 21 years old. Thus,

   $$x + x + 3 = 21$$
   $$2x + 3 = 21 \quad \text{Subtract 3 from each side}$$
   $$2x = 18 \quad \text{Divide each side by 2}$$
   $$x = 9$$

   Therefore, Joshua is 9 years old now.

7. (D)

   Tips Use the distributive property: $a(b + c) = ab + ac$

   $$\frac{x}{2} + \frac{x}{3} = 5 \quad \text{Multiply each side by 6}$$
   $$6\left(\frac{x}{2} + \frac{x}{3}\right) = 6(5) \quad \text{Use the distributive property}$$
   $$3x + 2x = 30$$
   $$x = 6$$

   Therefore, the value of x for which $\frac{x}{2} + \frac{x}{3} = 5$ is 6.

8. (H)

   2 is the smallest and only even prime number. Therefore, (H) is the correct answer.

9. (E)

Let x be the number. The product of 3 and the number and the sum of 3 and the number can be written as $3x$ and $x+3$, respectively. Thus,

$$3x = x + 3 \qquad \text{Subtract } x \text{ from each side}$$
$$2x = 3 \qquad \text{Divide each side by 2}$$
$$x = \frac{3}{2}$$

Therefore, the number is $\frac{3}{2}$.

10. (H)

Substitute 2 for x and evaluate the expression.

$$\begin{aligned} 2|x-5| - 3|1-x| &= 2|2-5| - 3|1-2| & \text{Substitute 2 for } x. \\ &= 2|-3| - 3|-1| & \text{Since } |-3| = 3 \text{ and } |-1| = 1 \\ &= 2(3) - 3(1) \\ &= 3 \end{aligned}$$

11. (A)

Let x be the positive integer. The positive integer has the remainder of 1 when it is divided by 3 and 4, respectively. This implies that one less than the positive integer, $x-1$, is divisible by 3 and 4; that is, $x-1$ is a least common multiple of 3 and 4, which is 12. The possible values of $x-1$ are $12, 24, 36, \cdots$. Thus, the possible values of x are $13, 25, 37, \cdots$. Therefore, (A) is the correct answer.

12. (G)

> Tips
> 1. The circumference of a circle with a radius of r: $C = 2\pi r$.
> 2. The area of a circle with a radius of r: $A = \pi r^2$.

Since the circumference and the area of the circle are the same, set them equal and solve for the radius, r.

$$\pi r^2 = 2\pi r \qquad \text{Divide each side by } \pi r$$
$$r = 2$$

Therefore, the radius of the circle is 2.

13. (A)

Since $-3 < x \leq 2$, the possible integer values of x are $-2, -1, 0, 1,$ and 2. Therefore, the sum of all possible integer values of x is $-2 + (-1) + 0 + 1 + 2 = 0$.

14. (K)

> Tips Use the exponent property: $\frac{x^n}{y^n} = \left(\frac{x}{y}\right)^n$

If $\frac{y}{x} = \frac{1}{2}$, $\frac{x}{y} = 2$. Therefore, the value of $\frac{x^2}{y^2} = \left(\frac{x}{y}\right)^2 = 2^2 = 4$.

SOLOMON ACADEMY

TEST 5 SOLUTIONS

15. (C)

$$3(x+y) = 12 \quad \text{Divide each side by 3}$$
$$x + y = 4 \quad \text{Subtract } y \text{ from each side}$$
$$x = 4 - y$$

Therefore, x in terms of y is $4 - y$.

16. (J)

Tips Counting integers = greatest integer − smallest integer + 1.

24 and 81 Inclusive means that 24 and 81 are included. Thus, the positive integers under consideration are 24, 25, $\cdots$, 80, and 81. Therefore, there are $81 - 24 + 1 = 58$ positive integers between 24 and 81 inclusive.

17. (E)

Let x be the number of coins that Jason has. The number of coins that Mr. Rhee has can be written as $2x$. Since Jason and Mr. Rhee have 60 coins together, $x + 2x = 60$. Thus, $x = 20$, which means that Jason has 20 coins. Since Joshua has half as many as coins as Jason, Joshua has 10 coins.

18. (K)

Tips Percent increase = $\dfrac{\text{new value} - \text{old value}}{\text{old value}} \times 100\%$

The price of a pencil increases from $1.20 to $1.68. The new value is $1.68 and old value is $1.20.

$$\text{Percent increase} = \frac{\$1.68 - \$1.20}{\$1.20} \times 100\%$$
$$= \frac{0.48}{1.20} \times 100\%$$
$$= 40\%$$

Therefore, the percent increase is 40%.

19. (C)

$$x^3 + 1 = 65 \quad \text{Subtract 1 from each side}$$
$$x^3 = 64 \quad \text{Since } 4^3 = 64$$
$$x = 4$$

Therefore, $x^2 + 1 = 4^2 + 1 = 17$.

20. (J)

Arrange the factors of 36 from least to greatest: 1, 2, 3, 4, 6, 9, 12, 18, 36. Since there are 9 factors of 36, the median is the 5th number, which is 6.

SOLOMON ACADEMY — TEST 5 SOLUTIONS

21. (D)

In the figure below, 4 congruent squares are placed side by side to form a rectangle. Let x be the side length of each square. Then, the length and width of the rectangle are $4x$ and x, respectively.

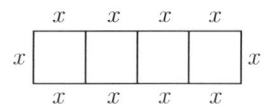

The area of the rectangle can be written as $4x \times x = 4x^2$. Since the area of the rectangle is 36, $4x^2 = 36$. Thus, $x = 3$. Therefore, the perimeter of the rectangle, $10x$, is $10(3) = 30$.

22. (F)

(Tips) If a number, which is equal to the mean of a set, is added to the set, the mean remains the same.

The mean of four numbers is 10 which means that the sum of these four numbers is $4 \times 10 = 40$. Since 10 is added to the sum of the four numbers, the sum of the five numbers is $40 + 10 = 50$. Therefore, the mean of the five numbers is $\frac{50}{5} = 10$.

23. (B)

There are 60 minutes in one hour. To find out how fast Mr. Rhee traveled in miles per hour, set up a proportion in terms of miles and minutes.

$$40\,\text{miles} : 40\,\text{minutes} = x\,\text{miles} : 60\,\text{minutes}$$

$$\frac{40}{40} = \frac{x}{60} \quad \text{Use cross product property}$$

$$40x = 40 \times 60 \quad \text{Divide each side by 40}$$

$$x = 60$$

$x = 60$ means that Mr. Rhee traveled 60 miles in one hour. Therefore, he traveled at 60 miles per hour.

24. (J)

$$xy + xz = 35 \quad \text{Substitute 3 for } y \text{ and 4 for } z$$
$$3x + 4x = 35 \quad \text{Simplify}$$
$$7x = 35 \quad \text{Divide each side by 7}$$
$$x = 5$$

Therefore, the value of x is 5.

25. (C)

(Tips) Use the distributive property: $(a + b)(c + d) = ac + ad + bc + bd$

In order to multiply two binomials, use the distributive property.

$$(x - 2)(x - 3) = x^2 - 3x - 2x + 6$$
$$= x^2 - 5x + 6$$

SOLOMON ACADEMY   Distribution or replication of any part of this page is prohibited.   TEST 5 SOLUTIONS

26. (F)

One-third of the students in a class are taking science classes. One half of the student taking science classes are taking chemistry class. This means that the fractional part of the students in the class who are chemisty is $\frac{1}{3} \times \frac{1}{2}$ or $\frac{1}{6}$.

27. (E)

In the figure below, $BA = 7$ and $CD = 8$. The two segments $\overline{BA}$ and $\overline{CD}$ are overlapped so that the four points A, C, B, and D are arranged in that order and are on the same line.

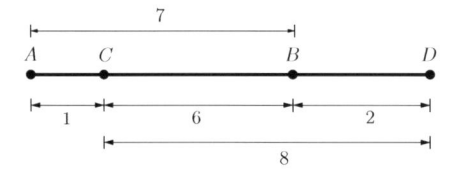

Since $BA = 7$ and $BC = 6$, $AC = 1$. Additionally, since $CD = 8$ and $BC = 6$, $BD = 2$. Therefore, the length of $\overline{AD}$ is $AC + BC + BD = 1 + 6 + 2$ or 9.

28. (J)

In the venn diagram below, define M, C, $M \cup C$, and $M \cap C$ as the number of students who are taking math, chemistry, either math or chemistry, and both math and chemistry, respectively.

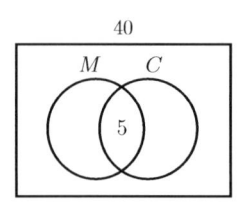

There are 40 students who are taking either math or chemistry and 5 students who are taking both math and chemistry. Thus, $M \cup C = 40$ and $M \cap C = 5$. Since the number of students taking math is twice as many as the number of students taking chemistry, $M = 2C$.

$$M \cup C = M + C - M \cap C \quad \text{Substitute } 2C \text{ for } M$$
$$40 = 2C + C - 5$$
$$3C = 45$$
$$C = 15$$

Since $M = 2C$, $M = 2 \times 15 = 30$. Therefore, the number of students taking math is 30.

29. (E)

Tips
1. Use the property of exponents: $a^x \cdot a^y = a^{x+y}$
2. $a^x \cdot a^y \cdot a^z = a^{x+y+z}$

$$3^n = 3^1 \times 3^2 \times 3^3 = 3^{1+2+3} = 3^6$$

Thus, $3^n = 3^6$. Therefore, $n = 6$.

SOLOMON ACADEMY · TEST 5 SOLUTIONS

30. (F)

$$\text{Base area} = \frac{\text{Volume of a cylinder}}{\text{Height}} = \frac{54\pi}{9} = 6\pi$$

31. (B)

$$\begin{aligned} 3(4x+6y) &= 3\cdot 2(2x+3y) &&\text{Factor out a 2 from inside the parenthesis} \\ &= 6(2x+3y) &&\text{Since } 2x+3y=5 \\ &= 30 \end{aligned}$$

Therefore, the value of $3(4x+6y)$ is 30.

32. (H)

Translate verbal phrases into mathematical expressions according to the guidelines shown below.

	Verbal phrase					
	what	percent	of	15	is	12
Math expression	x	$\frac{1}{100}$	$\times$	15	$=$	12

For instance, **what** can be translated into x, and **percent** can be translated into $\frac{1}{100}$. Furthermore, **of** can be translated into $\times$(multiplication), and **is** can be translated into $=$(equal sign). Thus, what percent of 15 is 12 can be translated to $x \times \frac{1}{100} \times 15 = 12$. Once the equation is set up, solve for x.

$$\begin{aligned} x \times \frac{1}{100} \times 15 &= 12 \\ \frac{15}{100}x &= 12 && \text{Multiply each side by } \frac{100}{15} \\ x &= 12 \times \frac{100}{15} \\ x &= 80 \end{aligned}$$

Therefore, (H) is the correct answer.

33. (C)

The degree measure of the arc in a circle is $360°$. If all six angles in the figure below were congruent, the measure of each angle would be $\frac{360}{6} = 60°$.

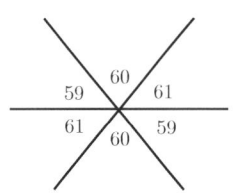

However, the measures of the six angles are integer values and not all congruent. Thus, the measure of some of the six angles are not $60°$. To determine the smallest possible degree measure of the largest angle, add the smallest positive integer value of the measure of an angle, $1°$, to $60°$. Therefore, the smallest possible degree measure of the largest angle is $61°$.

SOLOMON ACADEMY *Distribution or replication of any part of this page is prohibited.* **TEST 5 SOLUTIONS**

34. (H)

   In order to make a two-digit number, two numbers are selected at random from digits 1 through 4 without replacement. This means that the same number cannot be used for tens' and ones's place. Thus, the possible two-digit numbers are 12, 13, 14, 21, 23, 24, 31, 32, 34, 41, 42, and 43. Therefore, there are 12 two-digit numbers.

35. (E)

   As the value of x is increased by 1, the value of y is increased by 4.

x	2	3	4	5
y	3	3+4=7	7+4=11	11+4=k

   Therefore, the value of k when $x = 5$ is $11 + 4 = 15$.

36. (F)

   (Tips) The area of a trapezoid is $\frac{1}{2}(b_1 + b_2)h$, where b_1 and b_2 are the lengths of the top side and bottom side, respectively, and h is the height of the trapezoid.

   In the figure below, $b_1 = 4$, $b_2 = 8$, and $h = 5$.

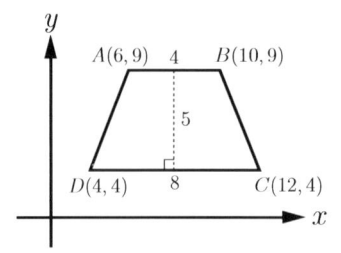

   Therefore, the area of the trapezoid is $\frac{1}{2}(4 + 8)(5) = 30$.

37. (D)

   Machine A can print 12 posters in 10 minutes or $12 \times 6 = 72$ posters in 60 minutes. Machine B can print 17 posters in 15 minutes or $17 \times 4 = 68$ posters in 60 minutes. In the first half hour, only machine A prints the posters. Thus, it prints $72 \times \frac{1}{2} = 36$ posters alone. In the next hour, both machine A and B print together. Thus, they print $72 + 68 = 140$ posters together. Therefore, the total number of posters that machine A and machine B print is $36 + 140 = 176$.

38. (G)

   (Tips) The sum of the measures of interior angles of a triangle is $180°$.

   Since $\frac{x}{2} + \frac{x}{6} = \frac{4x}{6} = \frac{2}{3}x$, $\frac{x}{2} + \frac{x}{6} + x = \frac{5}{3}x$.

   $$\frac{x}{2} + \frac{x}{6} + x = 180$$
   $$\frac{5}{3}x = 180$$
   $$x = 108$$

   Therefore, the measure of the smallest angle is $\frac{x}{6} = \frac{108}{6} = 18$.

162 www.solomonacademy.net

SOLOMON ACADEMY — Distribution or replication of any part of this page is prohibited. — TEST 5 SOLUTIONS

39. (A)

> **Tips** When two points (x_1, y_1) and (x_2, y_2) are given, the slope $= \frac{y_2 - y_1}{x_2 - x_1}$

The table below shows the three ordered pairs on a line: $(2, -1), (3, 2)$, and $(5, 8)$.

x	2	3	5
y	-1	2	8

Using two ordered pairs, $(3, 2)$ and $(5, 8)$, find the slope of the line.

$$\text{Slope} = \frac{y_2 - y_1}{x_2 - x_1} = \frac{8 - 2}{5 - 3} = 3$$

Therefore, the slope of the line is 3.

40. (G)

> **Tips** If n and k are prime numbers, the greatest common factor of n and k is 1.

List prime numbers: $2, 3, 5, 7, 11, \cdots$. It is worth noting that 2 is the smallest and the first prime number, and the only even prime number. Other than 2, all the prime numbers are odd numbers. If $n = 3$ and $k = 5$, $n + k = 8$ and $nk = 15$. Thus, eliminate answer choices (F) and (K). If $n = 2$ and $k = 3$, $n + k = 5$, which is not an even number and is not divisible by 2. Thus, eliminate answer choices (H) and (J). Since n and k are prime numbers, the greatest common factor of n and k is 1. Therefore, (G) is the correct answer.

41. (E)

> **Tips** Use the properties of exponents.
> 1. $a^x \cdot a^y = a^{x+y}$
> 2. $(a^x)^y = a^{xy}$

Since $8 = 2^3$ and $4^3 = (2^2)^3 = 2^6$,

$$8(4)^3 = 2^3 \cdot 2^6 = 2^{3+6} = 2^9$$

42. (G)

> **Tips** Use the difference of squares formula: $x^2 - y^2 = (x + y)(x - y)$

$$\begin{aligned} x^2 - y^2 &= (x + y)(x - y) \\ &= 10(4) \\ &= 40 \end{aligned}$$ Substitute 10 for $x + y$ and 4 for $x - y$

Therefore, the value of $x^2 - y^2$ is 40.

SOLOMON ACADEMY

Distribution or replication of any part of this page is prohibited.

TEST 5 SOLUTIONS

43. (C)

In the figure below, M, C, and B represent McLean, Centreville, and Brambleton, respectively. Mclean is 16 miles due north of Centreville. Brambleton is 12 miles due west of Centreville.

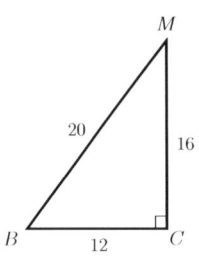

To find the distance between McLean and Brambleton, MB, use the Pythagorean theorem: $MB^2 = 12^2 + 16^2$. Thus, $MB = 20$. Therefore, McLean and Brambleton are 20 miles apart.

44. (K)

$\frac{2}{5}$ of the water tank is filled which means that $\frac{3}{5}$ of the water tank is empty. Set up a proportion as shown below.

$$1 \text{ minute} : \frac{2}{5} \text{ tank} = x \text{ minutes} : \frac{3}{5} \text{ tank}$$

$$\frac{1}{\frac{2}{5}} = \frac{x}{\frac{3}{5}} \qquad \text{Use cross product property}$$

$$\frac{2}{5}x = \frac{3}{5} \qquad \text{Multiply each side by } \frac{5}{2}$$

$$x = \frac{3}{2}$$

Therefore, it will take $\frac{3}{2}$ minute longer to fill the water tank.

45. (A)

The first and the second die have 6 possible outcomes each: 1, 2, 3, 4, 5, and 6, which are shown in the second row and the second column of the table below. There are total number of $6 \times 6 = 36$ possible outcomes. Each of the 36 outcomes represents the sum of the two numbers on the top of the first and the second die. For instance, when 2 is on the first die and 6 is on the second die, expressed as $(2, 6)$, the sum of the two numbers is 8.

		\multicolumn{6}{c}{1st die}
		1	2	3	4	5	6
2nd die	1						
	2						8
	3				8		
	4			8		10	
	5		8		10		
	6	8		10			

There are 8 outcomes for which the sum of the two numbers is either 8 or 10: $(2,6)$, $(3,5)$, $(4,4)$, $(5,3)$, $(6,2)$, and $(4,6)$, $(5,5)$, and $(6,4)$. Therefore, the probability that the sum of the two numbers on the top of each die is either 8 or 10 is $\frac{8}{36}$ or $\frac{2}{9}$.

164

www.solomonacademy.net

SOLOMON ACADEMY

Distribution or replication of any part of this page is prohibited.

TEST 5 SOLUTIONS

46. (J)

Translate the verbal phrases into mathematical expressions. Let x be the original number. A number is decreased by 3 and then the result is doubled can be expressed as $2(x-3)$. Six more than half the original number can be expressed as $\frac{x}{2}+6$. Set the two expressions equal to each other and solve for x.

$$2(x-3) = \frac{x}{2}+6 \qquad \text{Expand on the left side}$$
$$2x-6 = \frac{x}{2}+6 \qquad \text{Multiply each side by 2}$$
$$4x-12 = x+12$$
$$3x = 24$$
$$x = 8$$

Therefore, the value of the original number is 8.

47. (B)

A circular track whose circumference is 120 meters has 6 different labeled positions: A, B, C, D, E, and F. There are $\frac{120}{6}=20$ meters in between each different labeled position. Joshua and Jason run in opposite directions from position A at a rate of 4 meters per second clockwise and 2 meters per second counterclockwise, respectively. Let t be the time when Joshua and Jason meet for the first time. Then, the distances that Joshua and Jason run for time t are $4t$ and $2t$, respectively. When Joshua and Jason meet for the first time, the sum of the distances they run is equal to the circumference of the circular track, 120 meters. Set up an equation in terms of the sum of distances and solve for t.

$$4t + 2t = 120 \qquad \Longrightarrow \qquad t = 20$$

$t = 20$ implies that Joshua and Jason meet each other every 20 seconds. Joshua and Jason meet for the first time at $t = 20$ and second time at $t = 40$. At $t = 40$, Joshua runs $4 \times 40 = 160$ meters and Jason runs $2 \times 40 = 80$ meters. Therefore, Joshua and Jason meet each other for the second time at position C.

48. (K)

> **Tips** If two lines are parallel, their slopes are the same.

A line $ax + 3y = 4$ can be written in slope-intercept form as $y = -\frac{a}{3}x + \frac{4}{3}$. Since the two lines $ax + 3y = 4$ and $y = \frac{1}{2}x$ are parallel, their slopes must be the same. The slopes of the two lines are $-\frac{a}{3}$ and $\frac{1}{2}$. Thus,

$$-\frac{a}{3} = \frac{1}{2} \qquad \text{Multiply each side by } -3$$
$$a = -\frac{3}{2}$$

Therefore, the value of a is $-\frac{3}{2}$.

SOLOMON ACADEMY TEST 5 SOLUTIONS

49. (D)

> **Tips**   Distinct means different.

The mean of a set of five distinct positive integers is 8, which means that the sum of the five different positive integers is 40. Since the median of the set is 5, the set has the following five integers: p, q, 5, x, and y, where $p < q < 5 < x < y$. Since the sum of the five integers is 40,

$$\text{Sum} = p + q + 5 + x + y = 40$$

In order to find the greatest possible value of the largest integer, y, the smallest possible values of p, q, and x must be selected: $p = 1$, $q = 2$, and $x = 6$. Thus, $1 + 2 + 5 + 6 + y = 40$. Therefore, the greatest possible value of the largest integer, y, is 36.

50. (J)

> **Tips**   The surface area of a cube: $A = 6s^2$, where s is the length of a cube.

The length of cube A is increased by 25%, which means that it becomes 1.25 times longer than the original length. Since the length of cube A is 8, the length of cube B is $8 \times 1.25 = 10$.

$$\begin{aligned}\text{Difference of surface areas} &= \text{Surface area of cube } B - \text{Surface area of cube } A \\ &= 6(10)^2 - 6(8)^2 \\ &= 216\end{aligned}$$

Therefore, the surface area of cube B is 216 square inches greater than that of cube A.

SOLOMON ACADEMY   PRACTICE TEST 6

PRACTICE TEST 6
MATHEMATICS PROBLEMS
50 Questions
Time — 60 minutes

Directions: Solve each problem and enter your answer by marking the circle on the answer sheet. Choose the best answer among the answer choices given.

1. $14.8 \div 0.37 =$

    A. 400
    B. 40
    C. 4
    D. 0.4
    E. 0.04

2. What is 0.000107 written in scientific notation?

    F. 1.07×10^{-3}
    G. 10.7×10^{-3}
    H. 107×10^{-3}
    J. 10.7×10^{-4}
    K. 1.07×10^{-4}

3. If $5x + 1 = 26$, what is the value of x?

    A. 1
    B. 2
    C. 3
    D. 4
    E. 5

4. What is the largest prime factor of 51?

    F. 17
    G. 15
    H. 13
    J. 11
    K. 7

5. A set has the first n positive odd integers. If the sum of the all integers in the set is 36, what is the value of n?

    A. 3
    B. 4
    C. 5
    D. 6
    E. 7

6. Simplify $7x - 4y + 5 + 3y - 5x - 6$.

    F. $2x + y - 1$
    G. $2x - y - 1$
    H. $12x - y - 1$
    J. $12x - y + 11$
    K. $12x + 7y + 11$

167   www.solomonacademy.net

7. A sales tax rate is 10%. If Mr. Rhee pays $165 for a jacket including tax, what is the price of the jacket?

   A. $155
   B. $150
   C. $145
   D. $140
   E. $135

8. Which of the following number has the largest remainder if the number is divided by its units digit?

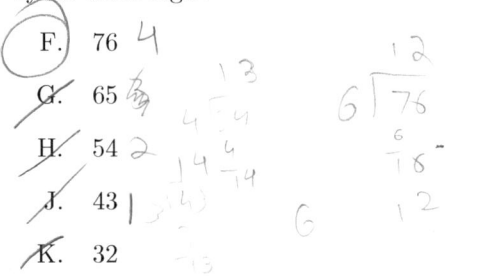

   F. 76
   G. 65
   H. 54
   J. 43
   K. 32

9. Evaluate $\frac{x+4}{4} + \frac{2x-3}{4} + \frac{3x-1}{4}$ when $x = 8$.

   A. 4
   B. 6
   C. 8
   D. 10
   E. 12

10. Joshua is seven years older than Jason. In seven years from now, the sum of their ages will be 25. What is the sum of their ages now?

    F. 9
    G. 11
    H. 14
    J. 18
    K. 21

11. If $0 < x \leq 10$ and $\frac{20}{x}$ is an integer, how many positive integer values of x are possible?

    A. 3
    B. 4
    C. 5
    D. 6
    E. 7

12. Two angles are complementary. If the measure of one angle is 30 less than the measure of the other angle, what is the measure of the smaller angle?

    F. 70°
    G. 60°
    H. 50°
    J. 40°
    K. 30°

13. If $2^x + 2^x + 2^x + 2^x = 2^6$, what is the value of x ?

    A. 4
    B. 5
    C. 6
    D. 7
    E. 8

14. Jason ran 1.8 miles on Monday, 2.4 miles on Tuesday, 2.5 miles on Wednesday, and 1.7 miles on Thursday. What is the average distance that Jason ran per day?

    F. 1.7 miles
    G. 1.9 miles
    H. 2.1 miles
    J. 2.3 miles
    K. 2.5 miles

15. Simplify $\dfrac{1}{\frac{1}{2}+\frac{1}{3}}$.

    A. $\frac{6}{5}$
    B. $\frac{5}{6}$
    C. $\frac{7}{6}$
    D. $\frac{1}{5}$
    E. 5

16. If $\dfrac{6x+9}{3} = y$, what is x in terms of y?

    F. $y+3$
    G. $\frac{1}{2}y - 3$
    H. $\frac{1}{2}(y-3)$
    J. $2y+3$
    K. $2(y-3)$

17. Three standard dice are rolled. What is the probability that the sum of the three numbers on top of each die is 18?

    A. $\frac{1}{6}$
    B. $\frac{1}{18}$
    C. $\frac{1}{36}$
    D. $\frac{1}{216}$
    E. $\frac{1}{300}$

18. If two concentric circles have radii of 2 and 3, what is the ratio of the area of the smaller circle to that of the larger circle?

    F. $1:2$
    G. $2:3$
    H. $3:5$
    J. $4:9$
    K. $5:12$

19. Jason has $10. If Joshua gives one-fourth of the amount of his money to Jason, both have the same amount of money. What is the amount of money that Joshua has in the beginning?

    A. $36
    B. $32
    C. $28
    D. $24
    E. $20

20. Evaluate $\sqrt{b^2 - 4ac}$ when $a = 3$, $b = -2$, and $c = -5$.

    F. 8
    G. 7
    H. 6
    J. 5
    K. 4

21. A set consists of prime numbers. If x is the smallest prime number in the set, what is the sixth smallest prime number in terms of x?

    A. $2x+1$
    B. $3x+1$
    C. $5x+1$
    D. $6x+1$
    E. $8x+1$

22. If Mr. Rhee drives 40 miles at 40 miles per hour and 20 miles at 60 miles per hour, what is Mr. Rhee's average speed, in miles per hour, for the entire tip?

    F. 52
    G. 50
    H. 48
    J. 45
    K. 42

23. What is the solution to $-2x + 5 > 7$?

    A. $x > -3$
    B. $x > -6$
    C. $x < -6$
    D. $x > -1$
    E. $x < -1$

$$\frac{1}{4}, \frac{1}{2}, 1, \ldots$$

24. In the sequence above, what is the value of the seventh term?

    F. 2
    G. 6
    H. 8
    J. 16
    K. 18

25. Joshua is running 2 feet per second. If Jason is 60 yards ahead of Joshua, how long will it take, in minutes and seconds, for Joshua to catch up to Jason?

    A. exactly 2 minutes
    B. 1 minute and 30 seconds
    C. exactly 1 minute
    D. 40 seconds
    E. 30 seconds

26. In the figure above, a rectangle whose area is 28 is divided into two smaller rectangles so that the ratio of the area of the larger rectangle to that of the smaller rectangle is 4 to 3. What is the area of the larger rectangle?

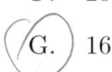

    F. 20
    G. 16
    H. 14
    J. 12
    K. 10

27. Evaluate $\frac{8(x-6)}{x-3}$ when $x = -1$.

    A. 10
    B. 12
    C. 14
    D. 16
    E. 18

28. If $0.2 < \dfrac{1}{x} < 0.5$, how many positive integer values of x are possible?

F. 5
G. 4
H. 3
J. 2
K. 1

x	1	2	3	4
y	5	5	5	5

29. In the table above, which of the following function contains the four ordered pairs shown on the table?

A. $x = 5$
B. $x = y - 4$
C. $y = x$
D. $y = x + 4$
E. $y = 5$

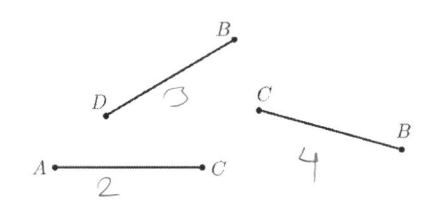

30. In the figure above, $AC = 2$, $BD = 3$, and $CB = 4$. If the three segments are put together so that points A, C, B, and D are arranged on the number line in that order, what is the length of segment $\overline{AD}$?

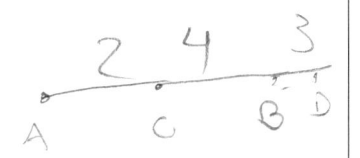

F. 10
G. 9
H. 8
J. 7
K. 6

31. In a triangle DEF, the measure of $\angle F$ is $90°$. If angle D and angle E are complementary and the measure of $\angle D$ is 30 more than twice the measure of $\angle E$, what is the measure of $\angle D$?

A. 60
B. 65
C. 70
D. 75
E. 80

32. If $x = 2y = 3z = 6$, what is the value of $x + y + z$?

F. 9
G. 10
H. 11
J. 12
K. 13

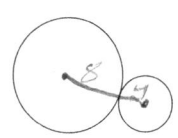

33. In the figure above, two circles are externally tangent to each other. The ratio of the radius of the larger circle to that of the smaller circle is 2 to 1. If the length of a segment connecting the center of each circle is 12, what is the circumference of the smaller circle?

A. 64π
B. 36π
C. 16π
D. 12π
E. 8π

34. One yard of wire is divided into two pieces. If the length of the shorter piece is $\frac{4}{5}$ of the length of the longer piece, what is the length of the longer piece in inches?

   F. 16 inches
   G. 18 inches
   H. 20 inches
   J. 22 inches
   K. 24 inches

35. Two stores A and B go on sale on the same item. Stores A and B give 12% and 21% discounts respectively. If the price of the item is $300 and you buy the item at store B, how much money would you save?

   A. $21
   B. $27
   C. $33
   D. $39
   E. $45

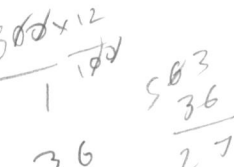

36. In the figure above, a circle is inscribed in a square. If the radius of the circle is 7, what is the perimeter of the square?

   F. 14
   G. 18
   H. 28
   J. 56
   K. 70

37. The positive difference of x and y is 17 and $x < y$. If $x = 18$, what is the value of y?

   A. 45
   B. 35
   C. 18
   D. 17
   E. 1

38. If $x < 0$, what is the value of x that satisfies $(x+1)^2 = 4$?

   F. 3
   G. 1
   H. -3
   J. -2
   K. -1

x	-1	1	2	4
y	10	4	1	k

39. The table above shows four points on a straight line. If the slope of the line is -3, what is the value of k?

   A. -1
   B. -2
   C. -3
   D. -4
   E. -5

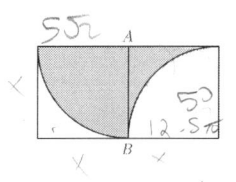

40. In the rectangle above, segment $\overline{AB}$ is drawn so that the rectangle is divided into two smaller squares. If the area of the rectangle is 100, what is the area of the shaded region?

F. 65
G. 60
H. 55
J. 50
K. 45

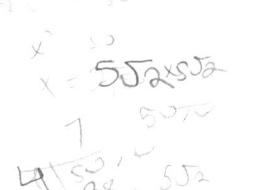

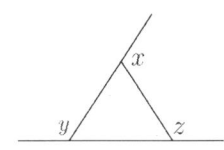

41. In the figure above, two sides of the triangle are expanded to form $\angle x$, $\angle y$, and $\angle z$. What is $m\angle x + m\angle y + m\angle z$?

A. 720
B. 380
C. 360
D. 180
E. 90

42. Mr. Rhee is running on a treadmill. He burns two calories every twenty seconds. How many calories will he burn if he runs $\frac{3}{4}$ hour on the treadmill?

F. 180 calories
G. 210 calories
H. 240 calories
J. 270 calories
K. 300 calories

43. Which of the following point below satisfies the equation $y + 2 = 3 - x$?

A. (3, 3)
B. (2, −2)
C. (1, 2)
D. (0, −1)
E. (−1, 2)

44. How many factors of 32 are even?

F. 6
G. 5
H. 4
J. 3
K. 2

45. The price of a pen is $3.60 in November and is increased by 100% in December. What is the price of the pen in December?

A. $7.20
B. $6.60
C. $6.00
D. $5.40
E. $4.80

46. A ferris wheel with a diameter of 60 feet takes 40 seconds to make a rotation. If Jason rides the ferris wheel for 1 minute and 40 seconds, through what angle, in degrees, does he rotate?

F. 360°
G. 684°
H. 720°
J. 810°
K. 900°

47. A ladder of 10 feet is leaning against the wall so that the bottom of the ladder is 6 feet away from the wall. If the top of the ladder starts sliding down the wall 1 foot per second, how far away is the bottom of the ladder from the wall after 2 seconds?

   A. 8 feet
   B. 7 feet
   C. 6 feet
   D. 5 feet
   E. 4 feet

48. There are four seats in a room and four students A, B, C, and D take each seat. If A and B do not want to sit at the end of either side, how many different seating arrangements are possible?

   F. 1
   G. 2
   H. 4
   J. 5
   K. 8

Year	...	3	5	7	...
Amount	...	$1950	$2450	$2950	...

49. The table above shows the amount of money that Sue has in her savings account over time. If the amount of money increases at a constant rate throughout the years, how much money did Sue deposit in her savings account in the beginning?

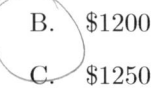

   A. $1150
   B. $1200
   C. $1250
   D. $1300
   E. $1350

50. The length, width, and height of a rectangular box is three, four, and five feet respectively. Each length, width, and height is divided by 3, 4, and 5 so that the rectangular box is divided into smaller cubes with sides of 1 foot. If one can of paint is needed to paint 12 square feet, how many cans of paint are needed to paint the surface area of all smaller cubes?

   F. 5
   G. 15
   H. 20
   J. 25
   K. 30

STOP

SOLOMON ACADEMY   TEST 6 SOLUTIONS

Answers and Solutions
Practice Test 6

Answers

1. B	2. K	3. E	4. F	5. D
6. G	7. B	8. F	9. E	10. G
11. C	12. K	13. A	14. H	15. A
16. H	17. D	18. J	19. E	20. F
21. D	22. J	23. E	24. J	25. B
26. G	27. C	28. J	29. E	30. G
31. C	32. H	33. E	34. H	35. B
36. J	37. B	38. H	39. E	40. J
41. C	42. J	43. E	44. G	45. A
46. K	47. A	48. H	49. B	50. K

Solutions

1. (B)
$$14.8 \div 0.37 = \frac{14.8 \times 100}{0.37 \times 100} = \frac{1480}{37} = 40$$

2. (K)

Tips: In scientific notation, all the numbers can be written in the form of $c \times 10^n$, where $1 \leq c < 10$ and n is an integer.

Since $0.1 = 10^{-1}$, $0.01 = 10^{-2}$, and $0.001 = 10^{-3}$, $0.0001 = 10^{-4}$.

$$0.000107 = 1.07 \times 0.0001 = 1.07 \times 10^{-4}$$

3. (E)
$$5x + 1 = 26 \quad \text{Subtract 1 from each side}$$
$$5x = 25 \quad \text{Divide each side by 5}$$
$$x = 5$$

4. (F)

The prime factorization of 51 is $51 = 3 \times 17$. Therefore, the largest prime factor of 51 is 17.

SOLOMON ACADEMY

Distribution or replication of any part of this page is prohibited.

TEST 6 SOLUTIONS

5. (D)

   Tips: The sum of the first n positive odd integers is n^2.

   $$1 = 1^2$$
   $$1 + 3 = 2^2$$
   $$1 + 3 + 5 = 3^2$$
   $$1 + 3 + 5 + 7 = 4^2$$
   $$1 + 3 + 5 + 7 + 9 = 5^2$$
   $$1 + 3 + 5 + 7 + 9 + 11 = 6^2$$

   Since the sum of the first six positive odd integers is 36, the value of n is 6.

6. (G)

   Group the like terms and simplify the expression.

   $$7x - 4y + 5 + 3y - 5x - 6 = (7x - 5x) + (-4y + 3y) + (5 - 6)$$
   $$= 2x - y - 1$$

7. (B)

   Let x be the price of the jacket. Since the sales tax is 10% of the price of the jacket, it can be expressed as $0.1x$. Thus, the amount of money that Mr. Rhee pays can be expressed as $x + 0.1x = 1.1x$. Thus,

   $$1.1x = 165 \qquad \text{Divide each side by 1.1}$$
   $$x = 150$$

   Therefore, the price of the jacket is $150.

8. (F)

   $$F.\ 76 \div 6 \implies r = 4$$
   $$G.\ 65 \div 5 \implies r = 0$$
   $$H.\ 54 \div 4 \implies r = 2$$
   $$J.\ 43 \div 3 \implies r = 1$$
   $$K.\ 32 \div 2 \implies r = 0$$

   Therefore, (F) is the correct answer.

9. (E)

   Since the three fractions, $\frac{x+4}{4}$, $\frac{2x-3}{4}$, and $\frac{3x-1}{4}$, have the same denominator, add the numerators. $\frac{x+4}{4} + \frac{2x-3}{4} + \frac{3x-1}{4} = \frac{6}{4}x$. Since $x = 8$, substitute 8 for x in $\frac{6}{4}x = \frac{6(8)}{4} = 12$. Therefore, the value of the expression is 12.

10. (G)

    Since the sum of Joshua's age and Jason's age in seven years from now is 25, the sum of their ages now is equal to $25 - 7 - 7 = 11$.

SOLOMON ACADEMY   Distribution or replication of any   TEST 6 SOLUTIONS
                      part of this page is prohibited.

11. (C)

   In order for $\frac{20}{x}$ to be an integer, x must be the factors of 20: 1, 2, 4, 5, 10, and 20. Since $0 < x \leq 10$, the possible positive integer values of x are 1, 2, 4, 5, and 10. Therefore, (C) is the correct answer.

12. (K)

   Tips: Complementary angles are two angles whose sum of their measures is 90°.

   Let x be the measure of the larger angle. Since the measure of the smaller angle is 30 less than the measure of the larger angle, it can be expressed as $x - 30$.

   $$x + x - 30 = 90 \quad \text{Since two angles are complementary}$$
   $$2x - 30 = 90 \quad \text{Add 30 to each side}$$
   $$2x = 120 \quad \text{Divide each side by 2}$$
   $$x = 60$$

   Therefore, the measure of the smaller angle is $x - 30 = 60 - 30 = 30°$.

13. (A)

   Tips:
   1. Use the property of exponents: $a^x \cdot a^y = a^{x+y}$.
   2. If the expressions have the same base, then the exponents on both sides are the same: If $a^x = a^y$, then $x = y$.

   Since you add 2^x four times, $2^x + 2^x + 2^x + 2^x = 4 \times 2^x$. Thus,

   $$2^x + 2^x + 2^x + 2^x = 4 \times 2^x$$
   $$= 2^2 \times 2^x$$
   $$= 2^{x+2}$$

   Since $2^{x+2} = 2^6$, $x + 2 = 6$. Thus, $x = 4$. Therefore, the value of x is 4.

14. (H)

   The total distance that Jason ran in four days is $1.8 + 2.4 + 2.5 + 1.7 = 8.4$ miles. Therefore, the average distance that Jason ran per day is $\frac{8.4}{4} = 2.1$ miles.

15. (A)

   Tips: $\dfrac{1}{\frac{1}{a} + \frac{1}{b}} \neq a + b$

   Simplify the two fractions in the denominator: $\frac{1}{2} + \frac{1}{3} = \frac{3}{6} + \frac{2}{6} = \frac{5}{6}$.

   $$\dfrac{1}{\frac{1}{2} + \frac{1}{3}} = \dfrac{1}{\frac{5}{6}} = \dfrac{6}{5}$$

177   www.solomonacademy.net

SOLOMON ACADEMY   Distribution or replication of any part of this page is prohibited.   TEST 6 SOLUTIONS

16. (H)

> **Tips** $\dfrac{a+b}{c} = \dfrac{a}{c} + \dfrac{b}{c}$.

Simplify the expression on the left side: $\dfrac{6x+9}{3} = \dfrac{6x}{3} + \dfrac{9}{3} = 2x + 3$. Thus,

$$\dfrac{6x+9}{3} = y \qquad \text{Simplify the expression on the left side}$$
$$2x + 3 = y \qquad \text{Subtract 3 from each side}$$
$$2x = y - 3 \qquad \text{Multiply each side by } \dfrac{1}{2}$$
$$x = \dfrac{1}{2}(y - 3)$$

Therefore, x in terms of y is $\dfrac{1}{2}(y - 3)$.

17. (D)

> **Tips** Probability(E) = $\dfrac{\text{The number of outcomes event E can happen}}{\text{The total number of possible outcomes}}$

Each die has six outcomes: 1 through 6. Since three dice are rolled, the total number of possible outcomes is $6 \times 6 \times 6 = 216$. In order to have the sum of the three numbers on top of each die is 18, the number on each die must be 6. This is only one outcome that gives the sum of 18. Therefore, the probability that the sum of the three numbers is 18 is $\dfrac{1}{216}$.

18. (J)

> **Tips** If the ratio of the radii of two circle is $a : b$, the ratio of the areas of the two circles is $a^2 : b^2$.

For simplicity, let the radii of the two circles be 2 and 3, respectively. Then, the areas of the smaller circle and the larger circle are $\pi(2)^2 = 4\pi$ and $\pi(3)^2 = 9\pi$. Therefore, the ratio of the area of the smaller circle to that of the larger circle is $4 : 9$.

19. (E)

Let x be the amount of money that Joshua has in the beginning. If Joshua gives one-fourth of the amount of his money to Jason, Jason has $\dfrac{x}{4} + 10$ and Joshua has $x - \dfrac{x}{4} = \dfrac{3x}{4}$. Since Joshua and Jason have the same amount of money,

$$\dfrac{3x}{4} = \dfrac{x}{4} + 10 \qquad \text{Multiply each side by 4}$$
$$3x = x + 40 \qquad \text{Subtract } x \text{ from each side}$$
$$2x = 40 \qquad \text{Divide each side by 2}$$
$$x = 20$$

Therefore, the amount of money that Joshua has in the beginning is $20.

20. (F)

Since $b^2 - 4ac = (-2)^2 - 4(3)(-5) = 64$, $\sqrt{b^2 - 4ac} = \sqrt{64} = 8$.

SOLOMON ACADEMY — TEST 6 SOLUTIONS

21. (D)

Prime numbers are 2, 3, 5, 7, 11, 13, $\cdots$. The smallest prime number is 2. Thus, $x = 2$. Since the six smallest prime number is 13, it can be expressed as $6x + 1$.

22. (J)

Tips: Average speed $= \dfrac{\text{Total distance}}{\text{Total time}}$

The total distance that Mr. Rhee drives is $40 + 20 = 60$ miles. It takes one hour to travel 40 miles at 40 miles per hour and $\frac{1}{3}$ hour to travel 20 miles at 60 miles per hour. Thus, the total time for the entire trip is $1 + \frac{1}{3} = \frac{4}{3}$ hour. Therefore, the average speed for the entire trip is $\dfrac{60 \text{ miles}}{\frac{4}{3} \text{ hour}} = 45$ miles per hour.

23. (E)

Tips: When multiplying or dividing an inequality by a negative number, reverse the inequality symbol.

$-2x + 5 > 7$ Subtract 5 from each side
$-2x > 2$ Divide each side by -2
$x < -1$ Reverse the inequality symbol

Therefore, the solution to $-2x + 5 > 7$ is $x < -1$.

24. (J)

This is a geometric sequence with the common ratio of 2

$$\frac{1}{4}, \frac{1}{2}, 1, 2, 4, 8, 16, \cdots$$

In other words, multiply each term by 2 to obtain the next term. Therefore, the value of the seventh term is 16.

25. (B)

There are three feet in one yard. Jason is $60 \times 3 = 180$ feet ahead of Joshua. If Joshua is running 2 feet per second, he can run 180 feet in 90 seconds, which is equal to one minutes and 30 seconds. Therefore, it will take one minute and 30 seconds for Joshua to catch up to Jason.

26. (G)

The ratio of the area of the larger rectangle to that of the smaller rectangles is $4:3$. Let $4x$ and $3x$ be the areas of the larger and smaller rectangles, respectively. Since the sum of the areas of the two smaller rectangles is 28, $4x + 3x = 28$ or $7x = 28$. Thus, $x = 4$. Therefore, the area of the larger rectangle is $4x = 4(4) = 16$.

SOLOMON ACADEMY

TEST 6 SOLUTIONS

27. (C)

To evaluate the expression, substitute -1 for x.

$$\frac{8(x-6)}{x-3} = \frac{8(-1-6)}{-1-3} \qquad \text{Substitute } -1 \text{ for } x$$
$$= \frac{8(-7)}{-4}$$
$$= 14$$

28. (J)

> **Tips** The inequality symbol is reversed when taking the reciprocal of each side of the inequality.

$$0.2 < \frac{1}{x} < 0.5 \qquad \text{Take the reciprocal of each side}$$
$$\frac{1}{0.2} > x > \frac{1}{0.5} \qquad \text{Inequality symbol is reversed}$$
$$2 < x < 5$$

Since x is a positive integer, the possible values of x for which $2 < x < 5$ is 3 and 4. Therefore, there are two possible integer values of x.

29. (E)

As the value of x changes, the value of y remains the same. Thus, the function that contains the four ordered pairs is $y = 5$. Therefore, (E) is the correct answer.

30. (G)

The three segments, $\overline{AC}$, $\overline{CB}$, and $\overline{BD}$, are put together so that points A, C, B, and D are arranged on the number line in that order as shown in the figure below.

Therefore, the length of segment $\overline{AD}$ is 9.

31. (C)

Let x be the measure of $\angle E$. Since the measure of $\angle D$ is 30 more than twice the measure of $\angle E$, the measure of $\angle D$ can be expressed as $2x + 30$. Angle D and angle E are complementary angles. Thus, the sum of their measures is $90°$.

$$2x + 30 + x = 90 \qquad \text{Simplify}$$
$$3x + 30 = 90 \qquad \text{Subtract 30 from each side}$$
$$3x = 60 \qquad \text{Divide each side by 3}$$
$$x = 20$$

Therefore, the measure of $\angle D$ is $2x + 30 = 2(20) + 30 = 70°$.

SOLOMON ACADEMY   TEST 6 SOLUTIONS

32. (H)

$x = 2y = 3z = 6$ means that $x = 6$, $2y = 6$, and $3z = 6$. Thus, $y = 3$ and $z = 2$. Therefore, the value of $x + y + z = 6 + 3 + 2 = 11$.

33. (E)

The ratio of the radius of the larger circle to that of the smaller circle is 2 to 1. Let x and $2x$ as the radii of the smaller circle and the larger circle, respectively.

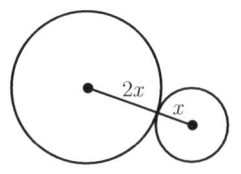

The length of the segment connecting the center of each circle can be expressed as $2x + x = 3x$ and equals 12. Thus, $3x = 12 \implies x = 4$. Therefore, the circumference of the smaller circle is $2\pi r = 2\pi(4) = 8\pi$.

34. (H)

There are 12 inches in one foot and 3 feet in one yard. Thus, there are 36 inches in one yard. Let $5x$ be the length of the longer piece. Since the length of the shorter piece is $\frac{4}{5}$ of the longer piece, the length of the shorter piece can be expressed as $\frac{4}{5} \times 5x = 4x$.

$$4x + 5x = 36 \qquad \text{Simplify}$$
$$9x = 36 \qquad \text{Divide each side by 9}$$
$$x = 4$$

Therefore, the length of the longer piece is $5x = 5(4) = 20$.

35. (B)

The price of the item is $300. Store A gives 12% discount and store B gives 21% discount on the same item. The difference of the discounts that store A and store B gives equals the amount of money you would save. Therefore, the amount of money that you would save is $21\% - 12\% = 9\%$ of $300 or $27.

36. (J)

Tips: If a circle is inscribed in a square, the length of the square is twice the radius of the circle.

The radius of the circle is 7. Since the circle is inscribed in the square, the length of the square is 14. Therefore, the perimeter of the square is $14 \times 4 = 56$.

37. (B)

$x < y$. The positive difference of x and y is 17 can be expressed as $y - x = 17$. Since $x = 18$, $y - 18 = 17$. Therefore, the value of y is 35.

38. (H)

   Take the square root of each side of the equation and solve for x.

   $$(x+1)^2 = 4$$
   $$x+1 = \pm 2$$
   $$x+1 = 2 \quad \text{or} \quad x+1 = -2$$
   $$x = 1 \quad \text{or} \quad x = -3$$

   Since $x < 0$, the only possible value of x is -3.

39. (E)

   > **Tips** When two points (x_1, y_1) and (x_2, y_2) are given, the slope $= \frac{y_2 - y_1}{x_2 - x_1}$

   Since the slope of the line is -3, choose two points $(2, 1)$ and $(4, k)$ to set up an equation in term of the slope.

   $$\frac{k-1}{4-2} = -3 \qquad \text{Since the slope is } -3$$
   $$\frac{k-1}{2} = -3 \qquad \text{Multiply each side by 2}$$
   $$k - 1 = -6 \qquad \text{Add 1 to each side}$$
   $$k = -5$$

   Therefore, the value of k is -5.

40. (J)

   Move the shaded region of the quarter circle that lies on the left square to the square on the right side as shown below.

   The area of the shaded region equals the area of the smaller square on the right side. Therefore, the area of the smaller square is half the area of the rectangle, or $\frac{1}{2}(100) = 50$.

41. (C)

   Angles x, y, and z are the exterior angles of the triangle. The sum of the exterior angles of any polygon is always $360°$. Therefore, $m\angle x + m\angle y + m\angle z = 360°$.

42. (J)

   There are three intervals of twenty seconds in one minute. Thus, Mr. Rhee burns $3 \times 2 = 6$ calories per minute when he is running on the treadmill. Since $\frac{3}{4}$ of an hour is $\frac{3}{4} \times 60 = 45$ minutes, Mr. Rhee will burn $45 \times 6 = 270$ calories if he runs on the treadmill.

SOLOMON ACADEMY Distribution or replication of any part of this page is prohibited. TEST 6 SOLUTIONS

43. (E)

Plug in the given values into the equation $y + 2 = 3 - x$ to see which is a solution.

(E) $(-1, 2)$ $y + 2 = 3 - x$ Substitute -1 for x and 2 for y
       $2 + 2 = 3 - (-1)$
       $4 = 4$ ✓ (Equation holds true)

Therefore, (E) is the correct answer.

44. (G)

$$\text{Factors of } 32 = \{1, 2, 4, 8, 16, 32\}$$

The even factors of 32 are 2, 4, 8, 16, and 32. Therefore, there are five even factors of 32.

45. (A)

The price of the pen is increased by 100%, which means that the price of the pen is doubled. Therefore, the price of the pen in December is $2 \times \$3.60 = \$7.20$.

46. (K)

There are 60 seconds in one minute. Convert 1 minute and 40 seconds to 100 seconds. Let x be the angle, in degrees, that the ferris wheel rotates for 100 seconds. Since the ferris wheel rotates $360°$ every 40 seconds, set up a proportion in terms of degrees (°) and seconds.

$$40_{\text{seconds}} : 360_{\text{degrees}} = 100_{\text{seconds}} : x_{\text{degrees}}$$
$$\frac{40}{360} = \frac{100}{x} \quad \text{Use cross product property}$$
$$40x = 100(360) \quad \text{Solve for } x$$
$$x = 900$$

Thus, the ferris wheel rotates $900°$ in one minute and 40 seconds, so does Jason. Therefore, Jason rotates $900°$ in one minute and 40 seconds.

47. (A)

The length of the ladder is 10 feet. Although the top and bottom of the ladder are sliding down and sliding away, the length of the ladder remains the same. When the bottom of the ladder is 6 feet away from the wall, the top of the ladder is 8 feet above the ground as shown in the diagram below. Use the Pythagorean theorem: $10^2 = a^2 + 6^2$, where a is the height of the wall at which the top of the ladder leans against. Thus, $a = 8$.

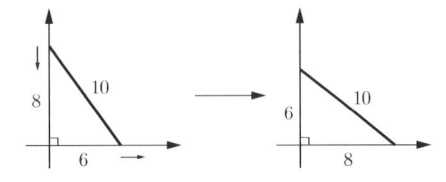

The top of the ladder is sliding down for 2 seconds at a rate of 1 foot per second so it is $8 - 2 = 6$ feet above the ground. To determine the distance between the bottom of the ladder and the wall, b, use the Pythagorean theorem: $10^2 = 6^2 + b^2$. Thus, $b = 8$. Therefore, the bottom of the ladder is 8 feet away from the wall after 2 seconds.

SOLOMON ACADEMY   Distribution or replication of any part of this page is prohibited.   TEST 6 SOLUTIONS

48. (H)

Since A and B do not want to sit at the end of either side, both C and D must sit at the end of either side. The possible seating arrangements are as follows:

$$C - A - B - D \quad D - A - B - C \quad C - B - A - D \quad D - B - A - C$$

Therefore, there are four possible seating arrangements.

49. (B)

The amount of money in Sue's savings account increases at a constant rate throughout the years. This suggests that a linear function best describes the information in the table. Create two ordered pairs, $(3, 1950)$ and $(5, 2450)$ from the table and find the slope of the linear function, which determines the rate at which the amount of money increases per year.

$$\text{Slope} = \frac{y_2 - y_1}{x_2 - x_1} = \frac{2450 - 1950}{5 - 3} = 250$$

Thus, the linear function can be written as $y = 250x + b$, where x represents the number of years, b represents the initial amount of money that Sue deposited in the beginning, and y represents the total amount of money in the savings account in x years. To find b, substitute 3 for x and 1950 for y.

$$y = 250x + b \quad \text{Substitute 3 for } x \text{ and 1950 for } y.$$
$$1950 = 250(3) + b \quad \text{Solve for } b$$
$$b = 1200$$

Therefore, Sue deposited $1200 in her savings account in the beginning.

50. (K)

In the figure below, each length, width, and height of the rectangular box is divided by 3, 4, and 5, respectively so that there are $3 \times 4 \times 5 = 60$ smaller cubes with sides of 1 foot.

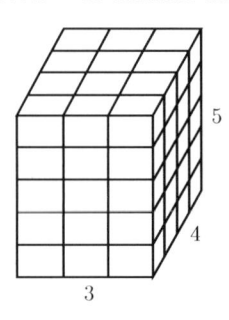

Each smaller cube has six faces. Thus, each has a surface area of 6 square feet. The total surface area of the 60 smaller cubes is $60 \times 6 = 360$ square feet. Since one can of paint is needed to paint 12 square feet, the total number of cans of paint is needed to paint the surface area of the 60 smaller cubes is $\frac{360 \text{ square feet}}{12 \text{ square feet/can}} = 30$ cans.

PRACTICE TEST 7
MATHEMATICS PROBLEMS
50 Questions
Time — 60 minutes

Directions: Solve each problem and enter your answer by marking the circle on the answer sheet. Choose the best answer among the answer choices given.

1. $\frac{3}{5} \times \frac{4}{3} \times \frac{2}{4} =$

    A. $\frac{5}{2}$

    B. $\frac{4}{3}$

    C. $\frac{3}{4}$

    D. $\frac{2}{5}$

    E. $\frac{1}{6}$

2. What is 25% of 24?

    F. 3

    G. 4

    H. 5

    J. 6

    K. 8

3. A box of twenty marbles contains two colors: red and blue. If there are three red marbles for every two blue marbles, how many red marbles are in the box?

    A. 6

    B. 8

    C. 10

    D. 12

    E. 14

4. Solve for x: $\frac{3}{2}x + 1 = 4$

    F. 6

    G. 5

    H. 4

    J. 3

    K. 2

5. If the circumference of a circle is 8π, what is the area of the circle?

    A. 4π

    B. 8π

    C. 10π

    D. 12π

    E. 16π

6. Simplify $2(x-3) - (2-3x)$.

    F. $-2x - 8$

    G. $-x - 1$

    H. $4x + 1$

    J. $5x - 8$

    K. $5x + 1$

7. If $72 = 2^a \times 3^b$, what is the sum of a and b?

   A. 4
   B. 5
   C. 6
   D. 7
   E. 8

8. Which of the following expression is equal to $-x(4-x)$?

   F. $-2x + 4$
   G. $-2x$
   H. $x^2 - 4x$
   J. $-x^2 - 4x$
   K. $4x^2 - x$

9. There are two black cards, three blue cards, and seven green cards in a bag. What is the probability that a non-green card is selected at random?

   A. $\frac{5}{12}$
   B. $\frac{7}{12}$
   C. $\frac{1}{3}$
   D. $\frac{1}{4}$
   E. $\frac{1}{6}$

10. If $p \blacktriangle q = p^q$ and $p \blacktriangledown q = q^p$, what is the value of $(3 \blacktriangle 1) \blacktriangledown 4$?

    F. 81
    G. 64
    H. 27
    J. 4
    K. 1

11. If the mean of three consecutive even integers is 48, what is the smallest integer?

    A. 42
    B. 44
    C. 46
    D. 48
    E. 50

12. If $x + 2y = 5$, what is $-x + 5$ in terms of y?

    F. $2y + 10$
    G. $2y$
    H. $2y - 5$
    J. $-2y$
    K. $-2y + 5$

13. Evaluate $\frac{1}{3}(x^2 - y^2)$ when $x = 27$ and $y = 24$.

    A. 43
    B. 45
    C. 47
    D. 49
    E. 51

14. If $x + y + 2z = 25$ and $x + y = 3z$, what is the value of z?

    F. 5
    G. 6
    H. 7
    J. 8
    K. 9

15. If decreasing x by 20% is the same as increasing y by 20%, what is the ratio of x to y?

    A. 1 : 1
    B. 1 : 2
    C. 2 : 3
    D. 3 : 2
    E. 4 : 3

$$3, 4, 5, 6, 0, 1, 2, 3, 4, \cdots$$

16. The numbers above show the remainders when consecutive integers are divided by an integer, n. What is the value of n?

    F. 3
    G. 4
    H. 5
    J. 6
    K. 7

17. If a number is divided by 2, then divided by 3, and then multiplied by 4, the result is 12. What is the number?

    A. 12
    B. 16
    C. 18
    D. 20
    E. 24

18. If x and y are distinct prime numbers, how many factors does the product of x and y have?

    F. 7
    G. 6
    H. 5
    J. 4
    K. 3

19. How many edges does a cube have?

    A. 6
    B. 8
    C. 9
    D. 10
    E. 12

20. If x is a negative integer and y is a positive integer, which of the following expression must be a positive integer?

    F. $y - x$
    G. $x + y$
    H. $y^2 - x^2$
    J. $x^2 - y^2$
    K. xy

21. In the sequence $2, -3, 4, -5, 6, -7, 8, \cdots$, what is the value of the 19th term?

    A. -20
    B. -19
    C. 19
    D. 20
    E. 21

x	-1	0	1
y	-2	3	8

22. The table above shows three ordered pairs on the graph of $y = mx + 3$. What is the value of m?

    F. 5
    G. 4
    H. 3
    J. 2
    K. 1

23. A circular shaped pizza has a radius of 6. If the pizza is cut into 6 equal slices, what is the sum of the area of the 4 slices?

    A. 32π
    B. 30π
    C. 28π
    D. 26π
    E. 24π

24. A number is an even integer. If the units digit of the number is divided by 2, how many digits from 0 to 9 will be possible for the units digit of the new number?

    F. 2
    G. 3
    H. 4
    J. 5
    K. 6

25. If the radius of a circle is $\dfrac{1}{\sqrt{\pi}}$, what is the area of the circle?

    A. 1
    B. 2
    C. 4
    D. π
    E. 4π

26. At a special sale event of a store, every three pens you purchase, you get the next two pens for free of charge. If the cost of one pen is $\$n$, what is the total cost of the 20 pens at the store?

    F. $\$10n$
    G. $\$12n$
    H. $\$20n$
    J. $\$10n + 10$
    K. $\$12n + 8$

27. Two angles A and B are supplementary angles. If the measure of angle B is 50% more than the measure of a right angle, what is the measure of angle A ?

    A. $60°$
    B. $55°$
    C. $50°$
    D. $45°$
    E. $40°$

28. If $7(x+3) = 35$, what is the value of $\dfrac{x+7}{3}$?

    F. 1
    G. 2
    H. 3
    J. 4
    K. 5

Month	Orange juice price
January	$3.59
February	$3.47
March	$3.51
April	$3.43
May	$3.56
June	$3.51

29. During what time period did the price of orange juice increase most?

   A. From January to February
   B. From February to March
   C. From March to April
   D. From April to May
   E. From May to June

30. If $\frac{8}{x-2} = \frac{12}{x-1}$, what is the value of x ?

   F. 5
   G. 4
   H. 3
   J. 2
   K. 1

31. If a number is divided by 2, 3, and 4, the remainders are 1, 2, and 1, respectively. Which of the following could be the number?

   A. 19
   B. 17
   C. 15
   D. 13
   E. 11

$$\frac{x+4}{3} + \frac{y+5}{3} + \frac{z+6}{3}$$

32. In the expression above, x, y, and z represent the measures of the interior angles of a triangle. What is the value of the expression above?

   F. 65°
   G. 70°
   H. 75°
   J. 80°
   K. 85°

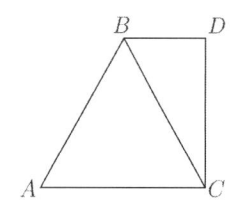

33. In the figure above, $ABDC$ is a trapezoid. If $m\angle C = 90°$ and $\triangle ABC$ is an equilateral triangle, what is the measure of $\angle ABD$?

   A. 100°
   B. 110°
   C. 120°
   D. 130°
   E. 140°

34. Which of the following expression is equal to $\frac{x^6}{x^2}$?

   F. $x \cdot x \cdot x$
   G. $(x^2)^3$
   H. $x^2 + x^2$
   J. $x^2 \cdot x^2$
   K. $(x^4)^{-1}$

35. Mr. Rhee went on a diet. If his weight was 180 pounds in August and 150 pounds in September, what percent of his weight did he lose?

   (A) $16\frac{2}{3}\%$
   (B) $33\frac{1}{3}\%$
   (C) 20%
   (D) 25%
   (E) 30%

36. There are two cubes. If the ratio of the volume of the smaller cube to that of the larger cube is 27 to 64, what is the ratio of the surface area of the smaller cube to that of the larger cube?

   F. 2 : 3
   G. 3 : 4
   H. 9 : 16
   J. 16 : 9
   K. 27 : 64

37. If the ratio of the interior angles of a triangle is 2 : 3 : 5, what is the degree measure of the largest angle?

   A. 36°
   B. 54°
   C. 90°
   D. 100°
   E. 120°

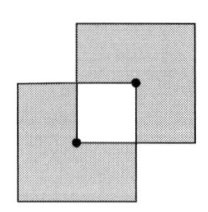

38. In the figure above, two identical squares are overlapped such that one vertex of each square is at the center of the other square. If the length of the square is 8, what is the area of the shaded region?

   F. 60
   G. 72
   H. 84
   J. 90
   K. 96

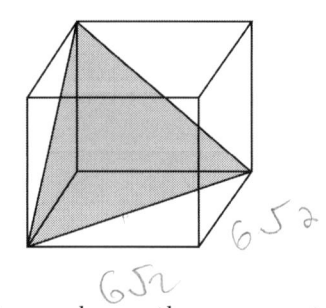

39. In the figure above, three segments are drawn to form a triangle. If the length of the cube is $6\sqrt{2}$, what is the perimeter of the triangle?

   A. 18
   B. 24
   C. $18\sqrt{2}$
   D. $24\sqrt{3}$
   E. 36

40. If the product of 250 and 4000 can be written as $1 \times 10^{n+1}$, what is the value of n?

    F. 4
    G. 5
    H. 6
    J. 7
    K. 8

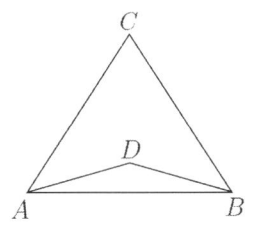

41. In the figure above, two triangles ACB and ADB are isosceles triangles. If $m\angle C = 84°$ and $m\angle DAB = \frac{1}{3}m\angle CAB$, what is the degree measure of $\angle D$?

    A. 84°
    B. 100°
    C. 116°
    D. 132°
    E. 148°

42. If $\left(\frac{1}{9}\right)^x = 4$, what is the value of 3^x?

    F. 3
    G. 2
    H. $\frac{1}{3}$
    J. $\frac{1}{2}$
    K. $\frac{4}{9}$

43. Using only the three digits, 1, 2, and 3, three-digit numbers are formed. If all the digits are used once, what is the probability that the three-digit numbers formed are divisible by 2?

    A. $\frac{1}{2}$
    B. $\frac{1}{3}$
    C. $\frac{1}{4}$
    D. $\frac{1}{5}$
    E. $\frac{1}{6}$

44. Mr. Rhee traveled 90 miles at 60 miles per hour to visit his parents. On the way home, he was caught in heavy traffic so that the average speed for the entire trip was 45 miles per hour. How fast did Mr. Rhee travel on the way home in miles per hour?

    F. 44 miles per hour
    G. 40 miles per hour
    H. 36 miles per hour
    J. 32 miles per hour
    K. 28 miles per hour

45. Joshua buys a package of three oranges for $2 and sells a package of two oranges for $3. If Joshua wants to make profit of $50, how many oranges does he need to sell?

    A. 40
    B. 50
    C. 60
    D. 70
    E. 80

46. Sue rented a car for two days. She paid the rental company a fixed daily fee plus an hourly charge for driving time. On the first day, she paid $89. On the second day, she drove the car twice as much as she did on the first day. So, she paid $139 on the second day. What is the fixed daily fee?

   F. $79
   G. $69
   H. $59
   J. $49
   K. $39

47. Jason filled a cup with equal amounts of orange juice, apple juice, and grape juice and mixed it well. He drank half of the cup and then only filled the cup with equal amounts of orange juice and apple juice. What fractional part of the cup was filled with the apple juice?

   A. $\frac{5}{12}$
   B. $\frac{1}{3}$
   C. $\frac{1}{4}$
   D. $\frac{1}{6}$
   E. $\frac{1}{12}$

$$S = 2 - 1 + 4 - 3 + 6 - 5 + \cdots + 50 - 49$$

48. If the pattern shown above continues, what is the value of S?

   F. 22
   G. 23
   H. 24
   J. 25
   K. 26

$$3x + 4y = 23$$
$$4x - 3y = 14$$

49. In the system of linear equations above, what is the value of $x + y$?

   A. 4
   B. 5
   C. 6
   D. 7
   E. 8

50. Jason spent one-third of the money in his savings account on books. A few days later, he spent half of the remaining money in his savings account on clothes. If Jason then had $175 left, how much money was in his savings account in the beginning?

   F. $225
   G. $375
   H. $525
   J. $675
   K. $825

SOLOMON ACADEMY — TEST 7 SOLUTIONS

Answers and Solutions
Practice Test 7

Answers

1. D	2. J	3. D	4. K	5. E
6. J	7. B	8. H	9. A	10. G
11. C	12. G	13. E	14. F	15. D
16. K	17. C	18. J	19. E	20. F
21. D	22. F	23. E	24. J	25. A
26. G	27. D	28. H	29. D	30. G
31. B	32. F	33. C	34. J	35. A
36. H	37. C	38. K	39. E	40. G
41. E	42. J	43. B	44. H	45. C
46. K	47. A	48. J	49. D	50. H

Solutions

1. (D)

$$\frac{3}{5} \times \frac{4}{3} \times \frac{2}{4} = \frac{2}{5}$$

2. (J)

   1% means 1 out of 100 or $\frac{1}{100}$. Thus, 25% means 25 out of 100, or $\frac{25}{100} = \frac{1}{4}$. Therefore, 25% of 24 is $\frac{1}{4} \times 24 = 6$.

3. (D)

   The ratio of the red marbles to blue marbles is 3 : 2. Let $3x$ and $2x$ be the numbers of the red marbles and blue marbles, respectively. Since the box contains 20 marbles, $3x + 2x = 20$ or $5x = 20$. Thus, $x = 4$. Therefore, the number of the red marbles is $3x = 3(4) = 12$.

4. (K)

$$\frac{3}{2}x + 1 = 4 \quad \text{Subtract 1 from each side}$$
$$\frac{3}{2}x = 3 \quad \text{Multiply each side by } \frac{2}{3}$$
$$x = 3 \times \frac{2}{3}$$
$$x = 2$$

   Therefore, the value of x is 2.

SOLOMON ACADEMY

Distribution or replication of any part of this page is prohibited.

TEST 7 SOLUTIONS

5. (E)

> **Tips**
> 1. Circumference of a circle with a radius of r is $C = 2\pi r$.
> 2. Area of a circle with a radius of r is $A = \pi r^2$.

Since the circumference of a circle is 8π, $C = 2\pi r = 8\pi$. Thus, the radius of the circle, r, is 4. Therefore, the area of the circle is $\pi r^2 = \pi(4)^2 = 16\pi$.

6. (J)

> **Tips** Use the distributive property: $a(b + c) = ab + ac$

$$\begin{aligned} 2(x-3) - (2-3x) &= 2x - 6 - 2 + 3x && \text{Use the distributive property} \\ &= (2x + 3x) + (-6 - 2) && \text{Simplify} \\ &= 5x - 8 \end{aligned}$$

7. (B)

$$\begin{aligned} 72 &= 2^a \times 3^b \\ 8 \times 9 &= 2^a \times 3^b \\ 2^3 \times 3^2 &= 2^a \times 3^b \end{aligned}$$

Thus, $a = 3$ and $b = 2$. Therefore, the sum of a and b is $a + b = 5$.

8. (H)

Use the distributive property: $a(b + c) = ab + ac$.

$$\begin{aligned} -x(4-x) &= -4x + x^2 && \text{Use the distributive property} \\ &= x^2 - 4x \end{aligned}$$

9. (A)

Out of the 12 cards, there are two black and three blue cards. Thus, there are 5 non-green cards. Therefore, the probability that a non-green card is selected at random is $\frac{5}{12}$.

10. (G)

According to the definition of $p▲q = p^q$, $3▲1 = 3^1 = 3$. Thus,

$$\begin{aligned} (3▲1)▼4 &= 3▼4 \\ &= 4^3 \\ &= 64 \end{aligned}$$

Therefore, the value of $(3▲1)▼4$ is 64.

SOLOMON ACADEMY   Distribution or replication of any   TEST 7 SOLUTIONS
                                   part of this page is prohibited.

11. (C)

> **Tips**
> 1. The difference between any two consecutive even integers is 2.
> 2. Sum of elements in a set = mean of elements × number of elements

Let x be the smallest integer. Then, $x+2$ and $x+4$ are the next two largest integers. Since the mean of the three consecutive even integers is 48, the sum of the three integers is $48 \times 3 = 144$.

$$x + x + 2 + x + 4 = 144 \qquad \text{Since the sum of the three integers is 144}$$
$$3x + 6 = 144 \qquad \text{Subtract 6 from each side}$$
$$3x = 138 \qquad \text{Divide each side by 3}$$
$$x = 46$$

Therefore, the smallest integer is 46.

12. (G)

$$x + 2y = 5 \qquad \text{Switch sides of the equation}$$
$$5 = x + 2y \qquad \text{Subtract } x \text{ from each side}$$
$$-x + 5 = 2y$$

Therefore, $-x+5$ in terms of y is $2y$.

13. (E)

> **Tips** Use the difference of squares formula: $x^2 - y^2 = (x+y)(x-y)$

In order to evaluate $27^2 - 24^2$ easily, use the difference of squares formula.

$$\frac{1}{3}(x^2 - y^2) = \frac{1}{3}(x+y)(x-y) \qquad \text{Substitute 27 for } x \text{ and 24 for } y$$
$$= \frac{1}{3}(27+24)(27-24)$$
$$= \frac{1}{3}(3)(51)$$
$$= 51$$

Therefore, $\frac{1}{3}(x^2 - y^2)$ when $x = 27$ and $y = 24$ is 51.

14. (F)

Substitute $3z$ for $x+y$ in the equation $x + y + 2z = 25$.

$$x + y + 2z = 25 \qquad \text{Substitute } 3z \text{ for } x+y$$
$$3z + 2z = 25 \qquad \text{Simplify}$$
$$5z = 25 \qquad \text{Divide each side by 5}$$
$$z = 5$$

Therefore, the value of z is 5.

15. (D)

20% means $\frac{20}{100} = \frac{1}{5}$. x is decreased by 20%. This can be expressed as $x - \frac{1}{5}x = \frac{4}{5}x$. Furthermore, y is increased by 20%. This can be expressed as $y + \frac{1}{5}y = \frac{6}{5}y$. Since $\frac{4}{5}x$ and $\frac{6}{5}y$ are the same, set these fractions equal and find the ratio of x to y.

$$\frac{4}{5}x = \frac{6}{5}y \qquad \text{Multiply each side by } \frac{5}{4}$$

$$x = \frac{5}{4} \times \frac{6}{5}y \qquad \text{Simplify}$$

$$x = \frac{3}{2}y \qquad \text{Divide each side by } y$$

$$\frac{x}{y} = \frac{3}{2}$$

Since $\frac{x}{y} = \frac{3}{2}$, the ratio of x to y is $3:2$.

16. (K)

Tips: If a number is divided by an integer n, the possible values of the remainder are $0, 1, 2, \cdots, n-1$.

The possible values of the remainder are 0, 1, 2, 3, 4, 5, and 6. This implies that the consecutive integers are divided by 7.

17. (C)

Let x be the number. Since the number is divided by 2 and then divided by 3, the number is divided by 6. This can be expressed as $\frac{x}{6}$. After multiplying $\frac{x}{6}$ by 4, the result is 12. Thus,

$$4 \times \frac{x}{6} = 12 \qquad \text{Simplify}$$

$$\frac{2}{3}x = 12 \qquad \text{Multiply each side by } \frac{3}{2}$$

$$x = \frac{3}{2}(12) \qquad \text{Simplify}$$

$$x = 18$$

Therefore, the number is 18.

18. (J)

Tips: If x and y are different prime numbers, the factors of xy are 1, x, y, and xy.

Let x be 2 and y be 3. The product of xy is 6. Thus, the factors of 6 are 1, 2, 3, and 6. Therefore, the number of factors of xy is 4.

SOLOMON ACADEMY — TEST 7 SOLUTIONS

19. (E)

A cube has 12 edges.

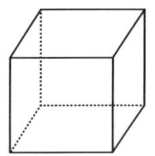

20. (F)

y is a positive integer and x is a negative integer. Since $y > 0$ and $x < 0$, $y - x > 0$. For instance, if $y = 2$ and $x = -1$, $y - x = 2 - (-1) = 3$. Therefore, (F) is the correct answer.

21. (D)

In the sequence $2, -3, 4, -5, 6, -7, 8, \cdots$, the 1st term is 2, the 3rd term is 4, the 5th term is 6, and the 7th term is 8. This pattern suggests that the 19th term is 20.

22. (F)

Select an ordered pair $(1, 8)$ that the graph $y = mx + 3$ passes through. Substitute 1 for x and 8 for y in $y = mx + 3$ to solve for m.

$$y = mx + 3 \quad \text{Substitute 1 for } x \text{ and 8 for } y$$
$$8 = m(1) + 3 \quad \text{Subtract 3 from each side}$$
$$m = 5$$

Therefore, the value of m is 5.

23. (E)

Tips   Area of a circle with a radius of r is πr^2.

A circular pizza with a radius of 6 is cut into 6 equal slices. In order to find the sum of the area of 4 slices, multiply the area of the entire pizza by $\frac{4}{6}$.

$$\text{Area of 4 slices} = \frac{4}{6} \times \pi(6)^2 = 24\pi$$

Therefore, the sum of the area of the 4 slices is 24π.

24. (J)

Since the number is an even integer, there are 5 possible units digits: 0, 2, 4, 6, and 8. If the number is divided by 2, possible units digits of the new number are 0, 1, 2, 3, and 4. Therefore, there are 5 possible units digits for the new number.

25. (A)

Tips
1. $(\sqrt{n})^2 = n$
2. $\left(\dfrac{a}{b}\right)^2 = \dfrac{a^2}{b^2}$

The radius of the circle is $\dfrac{1}{\sqrt{\pi}}$. Thus,

$$\text{Area of circle} = \pi\left(\dfrac{1}{\sqrt{\pi}}\right)^2 = \pi \times \dfrac{1}{\pi} = 1$$

Therefore, the area of the circle is 1.

26. (G)

Every three pens you purchase, you get the next two pens for free of charge. The cost of one pen is n. The cost of five pens is $3n$ since two pens are free. In other words, you pay $3n$ for every five pens you purchase. Therefore, the total costs for 20 pens is $4 \times \$3n = \$12n$.

27. (D)

Tips
1. Two angles are supplementary if the sum of the measures of the two angles is 180°.
2. 50% more than x can be expressed as $x + 0.5x$ or $1.5x$.

The measure of a right angle is 90°. The measure of angle B is 50% more than the measure of the right angle. Thus, the measure angle B is $1.5 \times 90° = 135°$. Since angles A and B are supplementary angles, the sum of their measures is 180°. Therefore, the measure of angle A is $180 - 135$ or 45°.

28. (H)

$$7(x+3) = 35 \quad \text{Divide each side by 7}$$
$$x + 3 = 5 \quad \text{Subtract 3 from each side}$$
$$x = 2$$

Therefore, the value of $\dfrac{x+7}{3}$ is $\dfrac{2+7}{3} = 3$.

29. (D)

There are two time periods at which the price of orange juice increases: from February to March and from April to May. From February to March, the price of orange juice is increased by $0.04. Furthermore, From April to May, the price of orange juice is increased by $0.13. Therefore, (D) is the correct answer.

SOLOMON ACADEMY — TEST 7 SOLUTIONS

30. (G)

Cross multiply and solve for x.

$$\frac{8}{x-2} = \frac{12}{x-1} \qquad \text{Cross multiply}$$
$$12(x-2) = 8(x-1) \qquad \text{Use the distributive property}$$
$$12x - 24 = 8x - 8 \qquad \text{Solve for } x$$
$$4x = 16 \qquad \text{Divide each side by 4}$$
$$x = 4$$

Therefore, the value of x is 4.

31. (B)

Instead of setting up an equation mathematically, use the given answer choices. 17 is the only number that satisfies the given conditions. If 17 is divided by 2, 3, and 4, the remainders are 1, 2, and 1, respectively.

32. (F)

Since x, y, and z represent the measures of the interior angles of a triangle, the sum of the measures of the interior angles expressed as $x + y + z$ is $180°$. Thus,

$$\frac{x+4}{3} + \frac{y+5}{3} + \frac{z+6}{3} = \frac{x+y+z+15}{3} \qquad \text{Substitute 180 for } x+y+z$$
$$= \frac{180+15}{3}$$
$$= 65$$

Therefore, the value of the expression is $65°$.

33. (C)

In the figure below, $ABDC$ is a trapezoid whose sides $\overline{BD} \parallel \overline{AC}$. Since $\overline{DC} \perp \overline{AC}$, $\overline{DC} \perp \overline{BD}$. Thus, $m\angle C = m\angle D = 90°$. $\triangle ABC$ is an equilateral triangle. Thus, $m\angle ABC = m\angle BCA = 60°$.

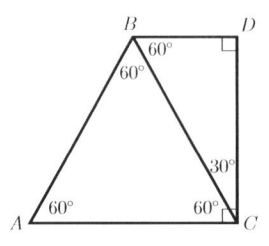

Since $m\angle C = 90°$ and $m\angle BCA = 60°$, $m\angle BCD = 30°$. $\angle DBC$ and $\angle BCD$ are complementary angles whose sum of their measures is $90°$. Thus, $m\angle DBC = 60°$. Since the measure of $\angle ABD$ equals the sum of the measures of $\angle ABC$ and $\angle DBC$, the measure of $\angle ABD$ is $60° + 60° = 120°$.

SOLOMON ACADEMY

Distribution or replication of any part of this page is prohibited.

TEST 7 SOLUTIONS

34. (J)

> **Tips** The properties of exponents.
> 1. $a^x \cdot a^y = a^{x+y}$
> 2. $\frac{a^x}{a^y} = a^{x-y}$
> 3. $(a^x)^y = a^{xy}$

According to the property of exponents, the expression $\frac{x^6}{x^2}$ is equal to $x^{6-2} = x^4$.

F. $x \cdot x \cdot x = x^3$

G. $(x^2)^3 = x^6$

H. $x^2 + x^2 = 2x^2$

J. $x^2 \cdot x^2 = x^4$

K. $(x^4)^{-1} = x^{-4}$

Therefore, (J) is the correct answer.

35. (A)

> **Tips** % decrease = $\frac{\text{Final value} - \text{Initial value}}{\text{Initial value}} \times 100\%$

Mr. Rhee's weight was 180 pounds in August and 150 pounds in September. Thus, the final value is 150 and initial value is 180.

$$\% \text{ decrease} = \frac{150 - 180}{180} \times 100\%$$
$$= -\frac{30}{180} \times 100\%$$
$$= -\frac{1}{6} \times 100\%$$
$$= -16\frac{2}{3}\%$$

Therefore, Mr. Rhee lost $16\frac{2}{3}\%$ of his weight.

36. (H)

For simplicity, let the volumes of the smaller cube and the larger cube be 27 and 64, respectively. Thus, the ratio of the volume of the smaller cube to that of the larger cube remains the same: $27 : 64$. The volume of a cube is s^3, where s is the side length of the cube. Thus, the side lengths of the smaller cube and the larger cube are 3 and 4, respectively. Since the surface area of a cube is $6s^2$, the ratio of the surface area of the smaller cube to that of the larger cube is $6(3)^2 : 6(4)^2$ or $9 : 16$.

SOLOMON ACADEMY — TEST 7 SOLUTIONS

37. (C)

The ratio of the interior angles of a triangle is $2:3:5$. Let $2x$, $3x$, and $5x$ be the measures of the interior angles of the triangle. Then, the sum of the measures of the interior angles can be expressed as $2x + 3x + 5x$ or $10x$. Since the sum of the measures of interior angles of any triangle is $180°$, $10x = 180°$. Thus, $x = 18°$. Therefore, the degree measure of the largest angle, $5x$, is $5 \times 18° = 90°$.

38. (K)

In the figure below, two identical squares with side length of 8 are overlapped such that one vertex of each square is at the center of the other square. The unshaded region represents the common area where the two squares overlap and is a square with side length of 4. Thus, the area of the unshaded region is 16.

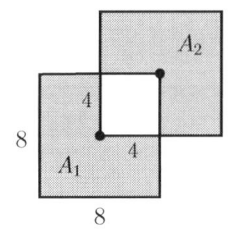

The area of the shaded region is the sum of the areas of A_1 and A_2. The area of each shaded region A_1 and A_2 is equal to the area of the square minus the area of the unshaded region, which is $64 - 16 = 48$. Therefore, the area of the shaded region is $A_1 + A_2 = 48 + 48 = 96$.

39. (E)

Each face of a cube is a square. In the figure below, the three segments $\overline{AC}$, $\overline{CD}$, and $\overline{AD}$ are the diagonals of the square faces and are same in length. Thus, $\triangle ACD$ is an equilateral triangle.

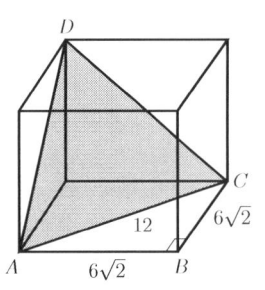

$\overline{AC}$ is the diagonal of the square face and is the hypotenuse of the $\triangle ABC$, which is a $45°$-$45°$-$90°$ special right triangle whose sides are in ratio $1:1:\sqrt{2}$. Since the length of the hypotenuse, AC, is $\sqrt{2}$ times the length of one leg, AB, $AC = AB \times \sqrt{2} = 6\sqrt{2} \times \sqrt{2} = 12$. Therefore, the perimeter of the triangle is $3 \times AC = 3 \times 12 = 36$.

40. (G)

Convert 250 and 4000 to scientific notation as 2.5×10^2 and 4×10^3, respectively.

$$250 \times 4000 = 2.5 \times 10^2 \times 4 \times 10^3$$
$$= 10 \times 10^5$$
$$= 1 \times 10^6$$

Since the product of 250 and 4000 can be written as $1 \times 10^{n+1}$, $n+1$ must be equal to 6. Therefore, the value of n is 5.

41. (E)

In the figure below, both $\triangle ACB$ and $\triangle ADB$ are isosceles triangles. The sum of the measures of interior angles of a triangle is $180°$. Since $m\angle C = 84°$, $m\angle CAB = m\angle CBA = \frac{180-84}{2} = 48°$.

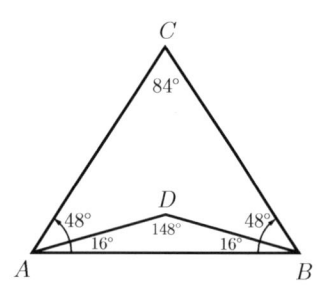

Since $m\angle DAB = \frac{1}{3}m\angle CAB$, $m\angle DAB = m\angle DBA = \frac{1}{3}(48°) = 16°$. Therefore, the degree measure of $\angle D$ is $180° - m\angle DAB - m\angle DBA$ or $180° - 16° - 16° = 148°$.

42. (J)

Tips
1. $\frac{1}{a} = a^{-1}$
2. $(a^x)^y = a^{xy}$

Since $\frac{1}{9} = \frac{1}{3^2} = 3^{-2}$, $\left(\frac{1}{9}\right)^x = (3^{-2})^x = 3^{-2x}$.

$$\left(\frac{1}{9}\right)^x = 4 \qquad \text{Since } \left(\frac{1}{9}\right)^x = 3^{-2x}$$
$$3^{-2x} = 2^2 \qquad \text{Raise each side to the power of } -\frac{1}{2}$$
$$\left(3^{-2x}\right)^{-\frac{1}{2}} = (2^2)^{-\frac{1}{2}}$$
$$3^x = 2^{-1}$$
$$3^x = \frac{1}{2}$$

Therefore, the value of 3^x is $\frac{1}{2}$.

SOLOMON ACADEMY — TEST 7 SOLUTIONS

43. (B)

If the digits, 1, 2, and 3, are used once, there are six three-digit numbers: 123, 132, 213, 231, 312, and 321. Out of six three-digit numbers, only 132 and 312 are divisible by 2. Therefore, the probability that the three-digit numbers formed are divisible by 2 is $\frac{2}{6}$ or $\frac{1}{3}$.

44. (H)

Mr. Rhee traveled 90 miles at 60 miles per hour to visit his parents. It took him $\frac{90\,\text{miles}}{60\,\text{mph}}$ or 1.5 hours to drive to his parents's house. The total distance for the entire trip is 2×90 miles or 180 miles. Since the average speed for the entire trip is 45 miles per hour, the total time for the entire trip is $\frac{180\,\text{miles}}{45\,\text{mph}}$ or 4 hours. This implies that it took Mr. Rhee $4 - 1.5 = 2.5$ hours to drive back to his home. Therefore, the rate at which Mr. Rhee traveled on the way home is $\frac{90\,\text{miles}}{2.5\,\text{hours}}$ or 36 miles per hour.

45. (C)

It's easier to calculate the profit if the number of oranges that Joshua buys and sells is the same. Since Joshua buys a package of 3 oranges and sells a package of 2 oranges, the least common multiple (LCM) of 3 and 2 is 6. This means that if Joshua buys 6 oranges for $2 \times \$2 = \$4$ and sells 6 oranges for $3 \times \$3 = \$9$, he makes $\$9 - \$4 = \$5$ profit for every 6 oranges that he sells. Therefore, the number of oranges that Joshua needs to sell to make profit of \$50 is 10×6 or 60.

46. (K)

Let x be the fixed daily fee and y be the total hourly charge on the first day. Sue paid \$89 which includes the fixed daily fee and the total hourly charge on the first day. This can be expressed as $x + y = 89$. On the second day, she drove the car twice as much as she did on the first day and paid \$139, which can be expressed as $x + 2y = 139$. Use the linear combinations method to solve for y.

$$x + 2y = 139$$
$$\underline{x + y = 89} \qquad \text{Subtract two equations}$$
$$y = 50$$

Since $y = 50$ and $x + y = 89$, $x = 39$. Therefore, the fixed daily fee, x, is \$39.

SOLOMON ACADEMY

Distribution or replication of any part of this page is prohibited.

TEST 7 SOLUTIONS

47. (A)

Jason filled a cup with equal amounts of orange juice, apple juice, and grape juice. Thus, the amount of each fruit juice was equal to $\frac{1}{3}$ of the cup as shown in figure 1.

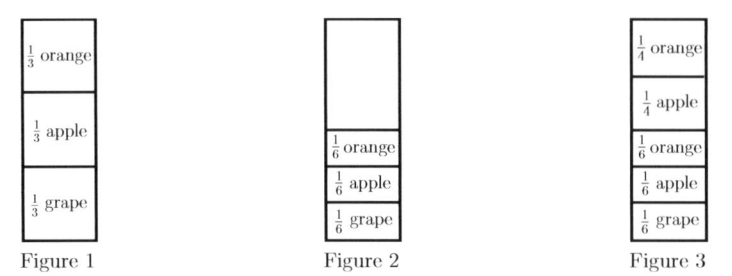

$\frac{1}{3}$ orange
$\frac{1}{3}$ apple
$\frac{1}{3}$ grape

Figure 1

$\frac{1}{6}$ orange
$\frac{1}{6}$ apple
$\frac{1}{6}$ grape

Figure 2

$\frac{1}{4}$ orange
$\frac{1}{4}$ apple
$\frac{1}{6}$ orange
$\frac{1}{6}$ apple
$\frac{1}{6}$ grape

Figure 3

After Jason drank half of the cup, the remaining fractional part of each fruit juice was $\frac{1}{6}$ of the cup as shown in figure 2. Afterwards, Jason filled the remaining cup with equal amounts of orange and apple juice. Since only half of the cup was available to fill, Jason added $\frac{1}{4}$ cup of orange juice and $\frac{1}{4}$ cup of apple juice as shown in figure 3. Therefore, the fractional part of the cup was filled with the apple juice is $\frac{1}{6} + \frac{1}{4} = \frac{5}{12}$.

48. (J)

Group 2 numbers to make a pair as shown below.

$$\begin{aligned} S &= 2 - 1 + 4 - 3 + 6 - 5 + \cdots + 50 - 49 \\ &= (2 - 1) + (4 - 3) + (6 - 5) + \cdots + (50 - 49) \\ &= 1 + 1 + 1 + \cdots + 1 \\ &= 25 \end{aligned}$$

The sum of each pair is 1. Since there are 25 pairs, the sum of the 25 pairs is 25. Therefore, the value of S is 25.

49. (D)

Use the linear combinations method. Multiply the first equation by 4 and the second equation by -3.

$$3x + 4y = 23 \xrightarrow{\text{Multiply by } 4} 12x + 16y = 92$$
$$4x - 3y = 14 \xrightarrow{\text{Multiply by } -3} -12x + 9y = -42$$

Add equations to eliminate x variables.

$$\begin{aligned} 12x + 16y &= 92 \\ -12x + 9y &= -42 \\ \hline 25y &= 50 \\ y &= 2 \end{aligned} \quad \text{Add two equations}$$

SOLOMON ACADEMY — TEST 7 SOLUTIONS

Substitute 2 for y in the second equation and solve for x.

$$4x - 3y = 14 \qquad \text{Substitute 2 for } y$$
$$4x - 3(2) = 14 \qquad \text{Add 6 to each side and then solve for } x$$
$$x = 5$$

Therefore, the value of $x + y = 5 + 2 = 7$.

50. (H)

Let x be the amount of money in Jason's savings account in the beginning. Jason spent one-third of his money, $\frac{x}{3}$, on books. The remaining money in his account is $x - \frac{x}{3}$ or $\frac{2}{3}x$. A few days later, Jason spent half of the remaining money, $\frac{1}{2} \times \frac{2}{3}x$ or $\frac{x}{3}$ on clothes. Thus, the remaining balance in his account after spending money on books and clothes is $x - \frac{x}{3} - \frac{x}{3}$ or $\frac{x}{3}$. The table below summarizes how much money Jason spent on books and clothes and the remaining balance in terms of x.

Initial amount	Books	Clothes	Remaining balance
x	$\frac{x}{3}$	$\frac{1}{2} \times \frac{2}{3}x = \frac{x}{3}$	$\frac{x}{3} = \$175$

Since the remaining balance, $\frac{x}{3}$, is \$175, solve the equation $\frac{x}{3} = 175$. Thus, $x = 525$. Therefore, Jason had \$525 in his savings account in the beginning.

PRACTICE TEST 8
MATHEMATICS PROBLEMS
50 Questions
Time — 60 minutes

Directions: Solve each problem and enter your answer by marking the circle on the answer sheet. Choose the best answer among the answer choices given.

1. $2.07 \div 0.23 =$
   A. 7
   B. 8
   C. 9
   D. 10
   E. 11

2. If the weight of a truck is 38500, what is 38500 written in scientific notation?
   F. 3.85×10^{-4}
   G. 38.5×10^{-3}
   H. 3.85×10^3
   J. 3.85×10^4
   K. 38.5×10^3

3. A water cooler was filled with water and then $\frac{7}{8}$ gallon of water was used. If $1\frac{7}{8}$ gallons of water is still remaining in the cooler, what is the capacity of the cooler?
   A. $2\frac{1}{2}$ gallons
   B. $2\frac{5}{8}$ gallons
   C. $2\frac{3}{4}$ gallons
   D. $2\frac{7}{8}$ gallons
   E. $3\frac{1}{8}$ gallons

4. What is the product of all factors of 9?
   F. 81
   G. 27
   H. 18
   J. 12
   K. 9

5. If $y = 4x + 5$ and $y = 25$, what is the value of x?
   A. 6
   B. 5
   C. 4
   D. 3
   E. 2

6. Evaluate $\sqrt{x^2 + 3x}$ when $x = -4$.
   F. 0
   G. 1
   H. 2
   J. 3
   K. 4

7. Jason mixes paint using 8 ounces of white paint for every 5 ounces of green paint. At this rate, how many ounces of green paint would be mixed with 32 ounces of white paint?

   A. 10
   B. 12
   C. 15
   D. 18
   E. 20

8. Which of the following table best represents the following verbal relationship? The number of boys, y, is three less than twice the number of girls, x.

   F.
x	2	3	4	5
y	4	7	10	13

   G.
x	2	3	4	5
y	-1	-7	-10	-13

   H.
x	2	3	4	5
y	-1	-3	-5	-7

   J.
x	2	3	4	5
y	1	-1	-3	-5

   K.
x	2	3	4	5
y	1	3	5	7

9. If $5^x = 125$, what is the value of 2^x ?

   A. 4
   B. 5
   C. 6
   D. 7
   E. 8

10. Which of the following equation is perpendicular to $y = -2x + 5$?

    F. $2x + y = 3$
    G. $2x - y = 3$
    H. $5x + y = 3$
    J. $x + 2y = 3$
    K. $x - 2y = 3$

11. The median of a set of six integers is 7. If the smallest integer is removed from the set, which of the following cannot be the new median?

    A. 6
    B. 7
    C. 8
    D. 10
    E. 11

12. Evaluate the expression $(x-3)(x+1)$ when $x = -2$.

    F. 6
    G. 5
    H. -1
    J. -5
    K. -6

13. Solve for x: $\dfrac{x}{5} + 2 = 5$

    A. 13
    B. 15
    C. 20
    D. 23
    E. 25

Quantity	5	10	20	40	80	...
Unit Price	$15	$14	$13	$12	$11	...

14. A t-shirt company provides lower prices if shirts are purchased in larger quantities. The table above shows their prices. If the amount of quantity reaches 320, what will be the unit price?

   F. $6
   G. $7
   H. $8
   J. $9
   K. $10

15. Joshua can wash 5 cars in one hour and Jason can wash 3 cars in one hour. At this rate, how long will it take Joshua and Jason to wash 26 cars if they work together?

   A. 4 hours and 15 minutes
   B. 3 hours and 45 minutes
   C. 3 hours and 30 minutes
   D. 3 hours and 15 minutes
   E. 2 hours and 45 minutes

16. If the diameter of a circle is 14, what is the ratio of the area of the circle to the circumference of the circle?

   F. 7 : 2
   G. 5 : 2
   H. 4 : 3
   J. 3 : 7
   K. 2 : 7

17. Triangle ABC is an obtuse triangle and is also an isosceles triangle. If the measure of the vertex angle of the triangle is 20% larger than the measure of a right angle, what is the measure of the base angle of the triangle?

   A. 36°
   B. 48°
   C. 64°
   D. 72°
   E. 90°

18. If the sum of scores on two math tests is 189 and the sum of scores on three English tests is 276, what is the average score of the five tests?

   F. 91
   G. 92
   H. 93
   J. 94
   K. 95

19. Simplify the expression $3(x+2)-2(5-3x)$.

   A. $-3x+4$
   B. $-3x-4$
   C. $4x+4$
   D. $9x+4$
   E. $9x-4$

20. Evaluate $\dfrac{x^6 \cdot x^4}{x^7}$ when $x=3$.

   F. 18
   G. 27
   H. 36
   J. 54
   K. 81

21. If $x = -2n+5$, $z = 4n+7$, and $x+2y = z$, what is y in terms of n ?

    A. $n+2$
    B. $n-2$
    C. $2n-2$
    D. $3n+1$
    E. $3n-1$

$$S = \frac{1+3+5+\cdots+49}{2+6+10+\cdots+98}$$

22. What is the value of the expression S shown above?

    F. $\frac{1}{2}$
    G. $\frac{2}{3}$
    H. $\frac{3}{4}$
    J. $\frac{5}{6}$
    K. $\frac{6}{7}$

23. What is the probability that a randomly selected positive integer less than 100 will be divisible by 7?

    A. $\frac{13}{99}$
    B. $\frac{14}{99}$
    C. $\frac{15}{99}$
    D. $\frac{13}{100}$
    E. $\frac{7}{50}$

24. Solve the inequality $-5x - 1 \leq 24$.

    F. $x < -5$
    G. $x \leq -5$
    H. $x \geq -5$
    J. $x \leq 5$
    K. $x > 5$

25. If A is a set of all multiples of 2 and B is a set of all multiples of 3, how many positive integer less than 15 belong to both A and B ?

    A. 0
    B. 1
    C. 2
    D. 3
    E. 4

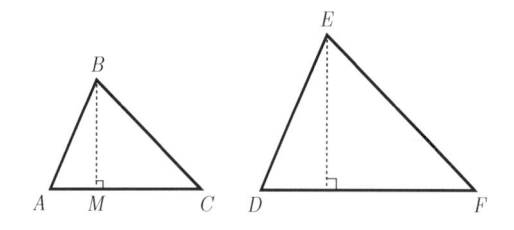

26. $\triangle ABC$ and $\triangle DEF$ are similar triangles. If $AC = 5$, $BM = 4$, and $DF = 15$, what is the area of $\triangle DEF$?

    F. 90
    G. 100
    H. 110
    J. 120
    K. 130

27. x, y, and z are positive integers such that $x < y < z$. Which of the following must be greatest?

    A. $y - x$
    B. $z - x$
    C. $y + x$
    D. $z + y$
    E. $z + x$

28. If ten more than twice a number equals four more than five times the number, what is the number?

    F. 1
    G. 2
    H. 3
    J. 4
    K. 5

29. $\sqrt{x-2} = 2$. What is the value of $|x+1|$?

    A. 3
    B. 4
    C. 5
    D. 6
    E. 7

30. In a high school football game, the blue team must move 15 yards forward to score a touchdown in the final four plays. The blue team moved 8 yards forward, 5 yards backward, and 4 yards backward for the first three plays. How many yards must the blue team move forward to win?

    F. 8
    G. 9
    H. 10
    J. 14
    K. 16

31. If $a = 4$, $b = -2$, $c = -1$, and $d = 3$, what is the value of $ac - bd$?

    A. 1
    B. 2
    C. 4
    D. 6
    E. 8

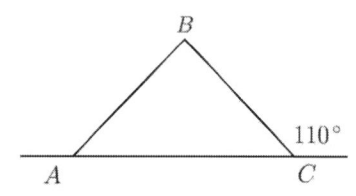

32. The degree measure of $\angle A$ is $\frac{2}{3}$ of the degree measure of $\angle B$. What is the degree measure of $\angle A$?

    F. 34
    G. 40
    H. 44
    J. 50
    K. 66

33. You are selling two types of tickets for an international culture night: student tickets and adult tickets. After selling 205 tickets, you realized that the number of student tickets sold is five less than four times as many as the number of adult tickets sold. What is the number of adult tickets sold?

    A. 32
    B. 36
    C. 42
    D. 85
    E. 163

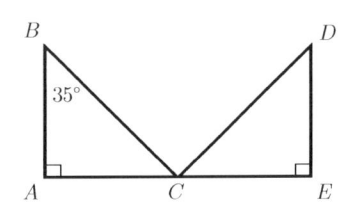

34. C is the midpoint of $\overline{AE}$. If $BC = DC$, which of the following is the degree measure of angle CDE?

F. 65
G. 55
H. 45
J. 35
K. 25

35. The coordinates of the three vertices of a right triangle are $(0,0)$, $(5,0)$, and $(5,12)$. What is the value of the length of the longest side of the triangle?

A. 13
B. 12
C. 11
D. 10
E. 9

36. A track team coach records the lap time after a student finishes his run. The student runs three times and his three lap times are 1 minutes and 45 seconds, 1 minutes and 55 seconds, and 2 minutes and 26 seconds. What is the average lap time for the three runs?

F. 1 minutes and 55 seconds
G. 1 minutes and 57 seconds
H. 1 minutes and 59 seconds
J. 2 minutes and 2 seconds
K. 2 minutes and 6 seconds

37. If $\dfrac{x}{y} = 5$, what is the value of $\dfrac{x+y}{y}$?

A. 6
B. 7
C. 8
D. 9
E. 10

38. Joshua has taken three tests in math class. The three test scores are 91, 86, and 87. In order for him to get an A in the class, he needs to get an average of at least 90 on the 4 tests. Which of the following score, s, must he receive to get an A? (Let s be the score on his 4th test.)

F. $s < 96$
G. $s \geq 96$
H. $s \leq 94$
J. $s > 94$
K. $s \geq 92$

39. The pie chart above shows the distribution of foreign languages for a group of high school students. If the number of students who take German is 42, what is the number of students who take French?

A. 55
B. 68
C. 75
D. 84
E. 114

40. Students walked into the Auditorium with rows of two. Jason noticed that his row is the 11th row from the front and the 10th row from the back. How many students went to the Auditorium?

   F. 20
   G. 21
   H. 40
   J. 80
   K. 82

41. If the radius of a circular garden is $\frac{100}{\pi}$ feet and a post is placed every four feet around the circumference of the garden, how many posts are there?

   A. 47
   B. 48
   C. 49
   D. 50
   E. 51

42. Five different points, A, B, E, C, and D lie on the same line in that order. The two points B and C trisect the segment $\overline{AD}$ whose length is 36. Point E is the midpoint of the segment $\overline{BC}$. What is the length of the segment $\overline{DE}$?

   F. 18
   G. 20
   H. 22
   J. 24
   K. 26

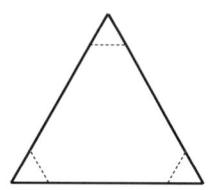

43. In the figure above, each corner of an equilateral triangle, with side length of 25, is cut into a smaller equilateral triangle with side length of 3. What is the perimeter of the remaining figure?

   A. 66
   B. 69
   C. 72
   D. 75
   E. 78

$$F = \frac{9}{5}C + 32$$

44. The equation above is used to convert degrees Celsius C to degrees Fahrenheit F. Which of the following must be the equation that convert degrees Fahrenheit F to degree Celsius C?

   F. $C = \frac{5}{9}(F + 32)$
   G. $C = \frac{5}{9}(F - 32)$
   H. $C = \frac{5}{9}F + 32$
   J. $C = \frac{5}{9}F - 32$
   K. $C = \frac{5F - 32}{9}$

45. Of the 43 students in a group, each student plays Tennis, Soccer, both or neither. 25 students play Tennis, 28 students play Soccer and 5 students play neither. If 15 students play both Tennis and Soccer, what is the sum of the number of students who play only Tennis or only Soccer?

    A. 16
    B. 23
    C. 33
    D. 38
    E. 43

46. Six people who work at the same rate painted half of a house in 4 days. How many additional people are needed to paint the remaining part of the house in 3 days?

    F. 6
    G. 5
    H. 4
    J. 3
    K. 2

47. You have a savings account. The amount A you have deposited into your account after t years can be modeled by a linear equation, $A = mt + b$. You have deposited $1175 into your savings account after three years and deposited $1850 after 6 years. How much money do you put into your savings account every year?

    A. $375
    B. $325
    C. $275
    D. $225
    E. $175

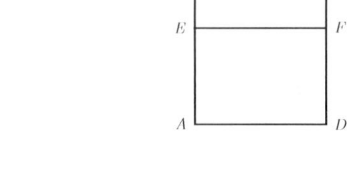

48. In the figure above, rectangle $ABCD$ is divided into two smaller rectangles. If the ratio of the area of rectangle $BCFE$ to that of rectangle $AEFD$ is 2 to 5, what is the ratio of BE to AB?

    F. $\frac{3}{4}$
    G. $\frac{2}{3}$
    H. $\frac{2}{5}$
    J. $\frac{2}{7}$
    K. $\frac{3}{10}$

$$xy = 1$$
$$x + \frac{1}{y} = 3$$

49. What is the value of $2x + 3y$?

   A. $\frac{1}{5}$
   B. $\frac{5}{2}$
   C. 5
   D. 6
   E. 9

50. When the positive integer k is divided by 5, the remainder is 1. When the positive integer n is divided by 7, the remainder is 2. Which of the following has a remainder of 2 when divided by 6?

   F. kn
   G. $kn + 1$
   H. $kn + 2$
   J. $kn + 3$
   K. $kn + 4$

SOLOMON ACADEMY — TEST 8 SOLUTIONS

Answers and Solutions
Practice Test 8

Answers

1. C	2. J	3. C	4. G	5. B
6. H	7. E	8. K	9. E	10. K
11. A	12. G	13. B	14. J	15. D
16. F	17. A	18. H	19. E	20. G
21. D	22. F	23. B	24. H	25. C
26. F	27. D	28. G	29. E	30. K
31. B	32. H	33. C	34. J	35. A
36. J	37. A	38. G	39. D	40. H
41. D	42. F	43. A	44. G	45. B
46. K	47. D	48. J	49. C	50. H

Solutions

1. (C)

$$2.07 \div 0.23 = \frac{2.07}{0.23} \times \frac{100}{100} = \frac{207}{23} = 9$$

2. (J)

$$38500 = 3.85 \times 10000 = 3.85 \times 10^4$$

3. (C)

After $\frac{7}{8}$ gallon of water was used, $1\frac{7}{8}$ gallon of water is still remaining in the cooler. Therefore, the capacity of the cooler is $\frac{7}{8} + 1\frac{7}{8} = 1\frac{14}{8} = 2\frac{3}{4}$ gallons.

4. (G)

The factors of 9 are 1, 3, and 9. Therefore, the product of all factors of 9 is $1 \times 3 \times 9 = 27$.

SOLOMON ACADEMY — TEST 8 SOLUTIONS

5. (B)

Substitute 25 for y in $y = 4x + 5$ and solve for x.

$$\begin{aligned} 4x + 5 &= y & &\text{Substitute 25 for } y \\ 4x + 5 &= 25 & &\text{Subtract 5 from each side} \\ 4x &= 20 & &\text{Divide each side by 4} \\ x &= 5 \end{aligned}$$

Therefore, the value of x is 5.

6. (H)

Substitute -4 for x in the expression $\sqrt{x^2 + 3x}$.

$$\begin{aligned} \sqrt{x^2 + 3x} &= \sqrt{(-4)^2 + 3(-4)} & &\text{Substitute } -4 \text{ for } x \\ &= \sqrt{4} \\ &= 2 \end{aligned}$$

Therefore, the value of $\sqrt{x^2 + 3x}$ when $x = -4$ is 2.

7. (E)

Let x be the number of ounces of green paint that would be mixed with 32 ounces of white paint. Set up a proportion and solve for x.

$$\begin{aligned} 8_{\text{white}} : 5_{\text{green}} &= 32_{\text{white}} : x_{\text{green}} \\ \frac{8}{5} &= \frac{32}{x} & &\text{Cross multiply} \\ 8x &= 32(5) & &\text{Divide each side by 8} \\ x &= 20 \end{aligned}$$

Therefore, 20 ounces of green paint would be mixed with 32 ounces of white paint.

8. (K)

The number of boys, y, is three less than twice the number of girls, x. This can be expressed as $y = 2x - 3$. The table in answer choice (K) best represents $y = 2x - 3$. Therefore, (K) is the correct answer.

9. (E)

Since $5^3 = 125$, $x = 3$. Therefore, the value of $2^x = 2^3 = 8$.

10. (K)

Tips: Perpendicular lines have negative reciprocal slopes.

The slope of $y = -2x + 5$ is -2. Thus, any equation that is perpendicular to $y = -2x + 5$ has a slope of $\frac{1}{2}$. Since $x - 2y = 3$ in answer choice (K) can be written as $y = \frac{1}{2}x - \frac{3}{2}$ and has a slope of $\frac{1}{2}$, (K) is the correct answer.

SOLOMON ACADEMY Distribution or replication of any part of this page is prohibited. TEST 8 SOLUTIONS

11. (A)

> **Tips**: If the smallest integer is removed from a set, the new median is greater than or equal to the previous median.

For example, set $A = \{2, 3, 6, 8, 10, 11\}$. The median is the average of the third and fourth number of the set, or 7. If the smallest integer, 2, is removed from the set, the new set $A = \{3, 6, 8, 10, 11\}$ and the new median is 8, which is larger than the previous median, 7. In other words, if the smallest integer is removed from the set, the new median cannot be smaller than the previous median. Therefore, (A) is the correct answer.

12. (G)

Substitute -2 for x in the expression.

$$(x-3)(x+1) = (-2-3)(-2+1) \qquad \text{Substitute } -2 \text{ for } x$$
$$= (-5)(-1)$$
$$= 5$$

Therefore, the value of $(x-3)(x+1)$ when $x = -2$ is 5.

13. (B)

$$\frac{x}{5} + 2 = 5 \qquad \text{Subtract 2 from each side}$$
$$\frac{x}{5} = 3 \qquad \text{Multiply each side by 5}$$
$$x = 15$$

Therefore, the value of x for which $\frac{x}{5} + 2 = 5$ is 15.

14. (J)

Observe the pattern in the table below. As the amount of quantity is doubled, the unit price is decreased by $1.

Quantity	5	10	20	40	80	160	320
Unit Price	$15	$14	$13	$12	$11	$10	$9

Therefore, when the amount of quantity reaches 320, the unit price is $9.

15. (D)

Joshua can wash 5 cars in one hour and Jason can wash 3 cars in one hour. This means that Joshua and Jason can wash 8 cars in one hour if they work together. Since $\frac{26 \text{ cars}}{8 \text{ cars per hour}} = 3.25$ hours, it will take Joshua and Jason 3.25 hours or 3 hours and 15 minutes to wash 26 cars.

16. (F)

If the diameter of a circle is 14, the radius of the circle is 7. Thus, the area of the circle is $\pi(7)^2 = 49\pi$ and the circumference of the circle is $2\pi(7) = 14\pi$. Therefore, the ratio of the area of the circle to the circumference of the circle is $49\pi : 14\pi$ or $7 : 2$.

SOLOMON ACADEMY   TEST 8 SOLUTIONS

17. (A)

> Tips
> 1. The sum of the measures of interior angles of a triangle is 180°.
> 2. In an isosceles triangle, the measures of the base angles are the same.

The measure of a right angle is 90°. The measure of the vertex angle is 20% larger than the measure of the right angle. Thus, it is $1.2 \times 90° = 108°$. Since triangle ABC is an isosceles triangle, the measures of the base angles are the same. Let x be the measure of the base angle of triangle ABC and set up an equation in terms of x.

$$x + x + 108 = 180 \quad \text{Since the sum of the measures of interior angles is } 180°$$
$$2x + 108 = 180 \quad \text{Subtract 108 from each side}$$
$$2x = 72 \quad \text{Divide each side by 2}$$
$$x = 36$$

Therefore, the measure of the base angle, x, is 36°.

18. (H)

Since the sum of scores on two math tests is 189 and the sum of scores on three English tests is 276, the sum of scores on five tests is $189 + 276 = 465$. Therefore, the average score of the five tests is $\frac{465}{5} = 93$.

19. (E)

$$3(x+2) - 2(5-3x) = 3x + 6 - 10 + 6x$$
$$= 3x + 6x + 6 - 10$$
$$= 9x - 4$$

20. (G)

> Tips
> 1. $a^x \cdot a^y = a^{x+y}$
> 2. $\dfrac{a^x}{a^y} = a^{x-y}$

Using the properties of exponents, $\dfrac{x^6 \cdot x^4}{x^7} = \dfrac{x^{10}}{x^7} = x^{10-7} = x^3$. Since $x = 3$, $x^3 = 3^3 = 27$.

21. (D)

Substitute $-2n + 5$ for x and $4n + 7$ for z in the equation $x + 2y = z$ and solve for y.

$$x + 2y = z \quad \text{Substitute } -2n+5 \text{ for } x \text{ and } 4n+7 \text{ for } z$$
$$-2n + 5 + 2y = 4n + 7 \quad \text{Add } 2n \text{ and then subtract 5 from each side}$$
$$2y = 6n + 2 \quad \text{Divide each side by 2}$$
$$y = 3n + 1$$

Therefore, y in terms of n is $3n + 1$.

SOLOMON ACADEMY — TEST 8 SOLUTIONS

22. (F)

Factor out 2 from the denominator and simplify the expression.

$$S = \frac{1+3+5+\cdots+49}{2+6+10+\cdots+98} = \frac{1+3+5+\cdots+49}{2(1+3+5+\cdots+49)} = \frac{1}{2}$$

Therefore, the value of expression S is $\frac{1}{2}$.

23. (B)

There are 99 positive integers less than 100: $1, 2, 3, \cdots, 98, 99$. Out of the 99 integers, there are 14 integers that are divisible by 7: $7, 14, 21, \cdots, 91, 98$. Therefore, the probability that a randomly selected positive integer less than 100 will be divisible by 7 is $\frac{14}{99}$.

24. (H)

$$-5x - 1 \leq 24 \qquad \text{Add 1 to each side}$$
$$-5x \leq 25 \qquad \text{Divide each side by } -5 \text{ and reverse the inequality symbol}$$
$$x \geq -5$$

Therefore, the solution to $-5x - 1 \leq 24$ is $x \geq -5$.

25. (C)

A is a set of all multiples of 2 and B is a set of all multiples of 3. It is possible to determine the answer by simply listing out both sets as shown below.

$$A = \{2, 4, 6, 8, 10, 12, 14, \cdots\}$$
$$B = \{3, 6, 9, 12, \cdots\}$$

Thus, there are only two positive integers less than 15 that belong to both sets A and B: 6 and 12. Additionally, if a number is both a multiple of 2 and 3, it is a multiple of 6. Thus, the positive integers less than 15 that are multiples of 6 are 6 and 12. Therefore, (C) is the correct answer.

26. (F)

Since the two triangles are similar triangles, set up a proportion to determine the height of triangle DEF. First, let's define x as the height of triangle DEF.

$$\frac{4}{x} = \frac{5}{15}$$
$$5x = 60$$
$$x = 12$$

Thus, the height of triangle DEF is 12.

$$\text{Area of } \triangle DEF = \frac{1}{2}bh$$
$$= \frac{1}{2}(15)(12)$$
$$= 90$$

Therefore, the area of $\triangle DEF$ is 90.

SOLOMON ACADEMY | Distribution or replication of any part of this page is prohibited. | TEST 8 SOLUTIONS

27. (D)

Plug in any positive integer values into the variables that match the description: x, y, and z are positive integers and z is greater than y which is greater than x. For instance, let $x = 1$, $y = 3$, and $z = 5$.

$$(A) \ y - x = 3 - 1 = 2$$
$$(B) \ z - x = 5 - 1 = 4$$
$$(C) \ y + x = 3 + 1 = 4$$
$$(D) \ z + y = 5 + 3 = 8$$
$$(E) \ z + x = 5 + 1 = 6$$

Thus, the sum of the two largest integers, $z + y$, is the greatest in value. Therefore, (D) is the correct answer.

28. (G)

Break each part of the verbal expression down. Let's define x as the number. Ten more than twice a number means $2x + 10$ and four more than five times the number means $5x + 4$. Set the two expressions equal to each other and solve for x.

$$5x + 4 = 2x + 10 \quad \text{Subtract } 2x \text{ from each side}$$
$$3x + 4 = 10 \quad \text{Subtract 4 from each side}$$
$$3x = 6 \quad \text{Divide each side by 3}$$
$$x = 2$$

Therefore, the number, x, is 2.

29. (E)

$$\sqrt{x - 2} = 2 \quad \text{Square each side}$$
$$x - 2 = 4 \quad \text{Add 2 to each side}$$
$$x = 6$$

Therefore, the value of $|x + 1|$ is $|6 + 1| = 7$.

30. (K)

The blue team needs to move the ball a total of 15 yards within 4 plays in order to win the football game. In the first three plays, the blue team moves the ball 8 yards forward, 5 yards backward, and 4 yards backward. The total gain in yards is $8 - 5 - 4 = -1$. Thus,

Number of yards on the 4th play $= 15 - (8 - 5 - 4) = 16$

Therefore, the blue team needs to move the ball 16 more yards forward in order to win the football game.

SOLOMON ACADEMY — TEST 8 SOLUTIONS

31. (B)

Plug in the given values into the expression $ac - bd$.

$$ac - bd = 4 \times -1 - (-2 \times 3)$$
$$= -4 - (-6)$$
$$= 2$$

Therefore, the value of $ac - bd$ is 2.

32. (H)

The sum of the measures of a triangle is 180°. Since the measure of angle BCA is 70°, the sum of the measures of angles A and B is 110°. Let x be the measure of angle B. Then, the measure of angle A is $\frac{2}{3}x$. Thus,

$$x + \frac{2}{3}x = 110 \qquad \text{Since } m\angle B + m\angle A = 110°$$
$$\frac{5x}{3} = 110 \qquad \text{Multiply each side by } \frac{3}{5}$$
$$x = 110 \times \frac{3}{5}$$
$$x = 66$$

Therefore, the measure of angle A, $\frac{2}{3}x$, is $\frac{2}{3} \times 66° = 44°$.

33. (C)

Let x be the number of adult tickets sold. Then, the number of student tickets sold can be expressed as $4x - 5$. Since the sum of the adult tickets and student tickets sold is 205 tickets, $x + 4x - 5 = 205$. Thus,

$$x + 4x - 5 = 205 \qquad \text{Simplify}$$
$$5x - 5 = 205 \qquad \text{Add 5 to each side}$$
$$5x = 210 \qquad \text{Divide each side by 5}$$
$$x = 42$$

Therefore, the number of adult tickets sold, x, is 42.

34. (J)

In the figure below, $BC = DC$ and $m\angle A = m\angle E = 90°$. Since C is the midpoint of $\overline{AE}$, $AC = EC$.

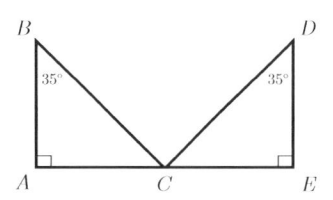

Thus, triangles CBA and CDE are congruent. Since the corresponding parts of the congruent triangles are congruent, $\angle CBA \cong \angle CDE$. Therefore, $m\angle CBA = m\angle CDE = 35°$.

SOLOMON ACADEMY   Distribution or replication of any part of this page is prohibited.   TEST 8 SOLUTIONS

35. (A)

In the figure below, the triangle is a 5-12-13 right triangle. Therefore, the length of the longest side, hypotenuse, is 13.

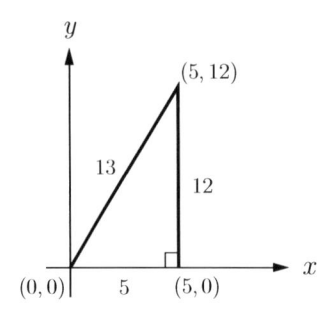

However, the answer can also be derived using the Pythagorean theorem. Define C as the length of the hypotenuse of the right triangle. Since C is the length, $C > 0$.

$$C^2 = 5^2 + 12^2$$
$$C = \sqrt{169}$$
$$C = 13$$

36. (J)

There are 60 seconds in 1 minute. Convert the time from minutes and seconds to just seconds.

$$1 \text{ minute and } 45 \text{ seconds} = 105 \text{ seconds}$$
$$1 \text{ minutes and } 55 \text{ seconds} = 115 \text{ seconds}$$
$$2 \text{ minutes and } 26 \text{ seconds} = 146 \text{ seconds}$$

Thus, the total lap time for the three runs is $105 + 115 + 146 = 366$ seconds.

$$\text{Average}_{\text{lap time}} = \frac{366 \text{ seconds}}{3 \text{ laps}}$$
$$= \frac{122 \text{ seconds}}{\text{lap}}$$

Therefore, the average lap time for the three runs is 122 seconds, which is 2 minutes and 2 seconds.

37. (A)

$$\boxed{\text{Tips}} \quad \frac{a+b}{c} = \frac{a}{c} + \frac{b}{c}$$

$$\frac{x+y}{y} = \frac{x}{y} + \frac{y}{y} \qquad\qquad \text{Since } \frac{x}{y} = 5$$
$$= 5 + 1$$
$$= 6$$

Therefore, the value of $\frac{x+y}{y}$ is 6.

222   www.solomonacademy.net

SOLOMON ACADEMY — TEST 8 SOLUTIONS

38. (G)

Joshua has taken three out of four tests and received the scores: 91, 86, and 87. In order to get an A for the math class, which is an average of 90 or more, the sum of the four tests must be at least $90 \times 4 = 360$. Set up the inequality and solve for s which represents the score needed on the 4th test.

$$91 + 86 + 87 + s \geq 360$$
$$s \geq 96$$

Joshua must obtain a score that is greater than or equal to 96 in order to receive an A for the math class.

39. (D)

Let's define x as the total number of students. Since 14% of the total number of students are taking German, the number of students taking German is $0.14x$. Thus,

$$0.14x = 42$$
$$x = \frac{42}{0.14} = 300$$

So, the total number of students is 300. Since 28% of the total number of students are taking French, the number of students taking French is $0.28 \times 300 = 84$.

40. (H)

Since Jason's row is the 11th row from the front, there are 10 rows in front of Jason. Since Jason's row is the 10th row from the back, there are 9 rows behind Jason. Thus, there are $10 + 1 + 9 = 20$ rows in the Auditorium. Since each row is occupied by 2 students, there are $20 \times 2 = 40$ students who went to the Auditorium.

41. (D)

The circumference of a circle with a radius of $\frac{100}{\pi}$ is $2\pi(\frac{100}{\pi}) = 200$. Thus,

$$\text{Total number of posts on a circle} = \frac{\text{Circumference of a circle}}{\text{Distance between each post}} = \frac{200}{4} = 50$$

Therefore, the total number of posts around the circular garden is 50.

42. (F)

In the figure below, $AB = BC = DC = 12$ and $CE = 6$.

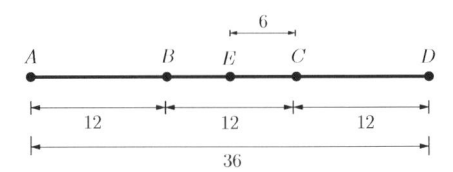

Therefore, $DE = DC + CE = 12 + 6 = 18$.

223

SOLOMON ACADEMY   Distribution or replication of any part of this page is prohibited.   TEST 8 SOLUTIONS

43. (A)

In the figure below, each side of the equilateral triangle in cut into a smaller equilateral triangle with side length of 3. So, the remaining figure is a hexagon.

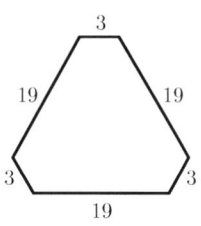

```
          3
       /     \
     19       19
     /         \
    3           3
       \     /
         19
```

Therefore, the perimeter of the hexagon is $19 \times 3 + 3 \times 3 = 66$.

44. (G)

$$\frac{9}{5}C + 32 = F \qquad \text{Subtract 32 from each side}$$

$$\frac{9}{5}C = F - 32 \qquad \text{Multiply each side by } \frac{5}{9}$$

$$C = \frac{5}{9}(F - 32)$$

45. (B)

The venn diagram below shows the complete breakdown of the given information.

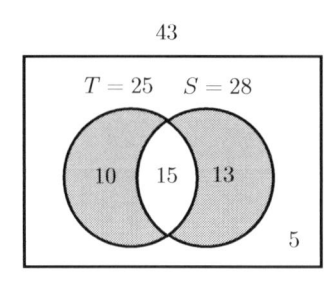

```
            43
   ┌─────────────────┐
   │  T = 25  S = 28 │
   │   ___   ___     │
   │  / 10\ /13 \    │
   │ (    15    )   │
   │  \___/ \___/    │
   │              5  │
   └─────────────────┘
```

The shaded region in the venn diagram represents the sum of the number of students who play only Tennis or only Soccer, which is $10 + 13 = 23$.

46. (K)

If it takes 6 people 4 days to paint half a house, 6 people will need 8 days to paint an entire house. In other words, the amount of work required to paint the entire house is equivalent to 6 $_{\text{people}} \times 8$ $_{\text{days}} = 48$ $_{\text{people} \times \text{days}}$. Since half the house remains unpainted, the amount of work required is $\frac{1}{2} \times 48$ $_{\text{people} \times \text{days}} = 24$ $_{\text{people} \times \text{days}}$. Let x be the total number of people needed to paint half the house in 3 days. Then, the amount of work required in 3 days is x $_{\text{people}} \times 3$ $_{\text{days}} = 3x$ $_{\text{people} \times \text{days}}$. Set up an equation in terms of the amount of work required and solve for x.

$$3x \text{ }_{\text{people} \times \text{days}} = 24 \text{ }_{\text{people} \times \text{days}}$$
$$x = 8$$

Thus, 8 people are needed to paint half the house in 3 days. Since there are already 6 people working on the house, the number of additional people needed is $8 - 6 = 2$ people.

47. (D)

In 3 years, you deposited $1175 and in 6 years, you deposited $1850. This can be written as ordered pairs, (x,y), where x represent years and y represents the amount deposited: $(3,1175)$ and $(6,1850)$. Find the slope of the line that passes through the points, $(3,1175)$ and $(6,1850)$, because it represents the amount of money deposited into the savings account every year.

$$\text{Slope} = \frac{y_2 - y_1}{x_2 - x_1}$$
$$= \frac{1850 - 1175}{6 - 3}$$
$$= \frac{675}{3}$$
$$= 225$$

Therefore, the amount of money deposited into the savings account every year is $225.

48. (J)

Define y as the length of the two smaller rectangles, $BCFE$ and $EFDA$. The areas of the two smaller rectangles $BCFE$ and $AEFD$ are $y \times BE$ and $y \times AE$, respectively. Since the ratio of areas of the two smaller rectangles is $2:5$, set up a proportion and find the ratio of BE to AE.

$$\text{Area of BCFE : Area of AEFD} = 2:5$$
$$y \times BE : y \times AE = 2:5$$
$$\frac{y \times BE}{y \times AE} = \frac{2}{5} \qquad y \text{ cancels each other}$$
$$\frac{BE}{AE} = \frac{2}{5}$$

Thus, the ratio of BE to AE is $2:5$. For simplicity, let $BE = 2$ and $AE = 5$. Since $AB = AE + BE = 7$, the ratio of BE to AB is $\frac{2}{7}$.

49. (C)

Tips: Use the distributive property: $a(b+c) = ab + ac$.

$$x + \frac{1}{y} = 3 \qquad \text{Multiply each side by } y$$
$$y\left(x + \frac{1}{y}\right) = 3(y) \qquad \text{Use the distributive property}$$
$$xy + 1 = 3y \qquad \text{Since } xy = 1$$
$$1 + 1 = 3y \qquad \text{Divide each side by 3}$$
$$y = \frac{2}{3}$$

Since $y = \frac{2}{3}$ and $xy = 1$, $x = \frac{3}{2}$. Therefore, the value of $2x + 3y$ is $2(\frac{3}{2}) + 3(\frac{2}{3}) = 5$.

SOLOMON ACADEMY

TEST 8 SOLUTIONS

50. (H)

The fastest way to solve this particular problem is to plug in numbers and observe which set of numbers satisfies the requirements in the problem. For instance, $k = 6$ and $n = 9$.

$$(A) \quad kn = 54 \implies \text{remainder of } 0$$
$$(B) \quad kn + 1 = 55 \implies \text{remainder of } 1$$
$$(C) \quad kn + 2 = 56 \implies \text{remainder of } 2$$
$$(D) \quad kn + 3 = 57 \implies \text{remainder of } 3$$
$$(E) \quad kn + 4 = 58 \implies \text{remainder of } 4$$

Made in the USA
Lexington, KY
04 November 2014